좌향, 여백, 표층
Prospect, Void, Surface

차례

에디토리얼		
Editorial	5	남겨진 유형의 조각들: 상업가로의 경우
		What Remains of What Was There: The Case of the Commercial Streets
		정평진, PyeongJin Jeong
전략		
Strategies	13	비틀어지는 도시의 좌향
		Twisting and Turning: Reorientating Toward the City
		버려지지 않는 공동의 여백
		The Common Void: Space Beyond Properties
		소비되는 표층의 두터움
		Consumed Yet Tangible: Thick Surfaces
에세이		
Essay	109	유형과 체계의 실험과 한계, 그리고 다른 가능성
		Reconfiguring Typology and Systems of Practice
		현명석, MyeonSeok Hyun
부록		
Appendix	127	리슈 건축 프로젝트 2012–2020
		Richue Architecure's Works 2012–2020

에디토리얼

남겨진 유형의 조각들: 상업가로의 경우

정평진[*]

Editorial

What Remains of What Was There: The Case of the Commercial Streets

Pyung Jin Jeong[*]

*
건축 잡지에서 기자로 일했고, 여러 매체에 건축과 관련된 글을 썼다. 서울시립대학교에서 건축을 전공했다.

PyeonJin Jeong worked as a editor for an architectural magazines and wrote articles in various media. He majored in architecture at University of Seoul

잦은 변화가 일어나는 도시에서 지속적인 생명력을 발휘하는 건축의 유형은 좀처럼 찾아보기 어렵다. 빠른 유속으로 퇴적이 이뤄지지 않는 강의 상류처럼 장소의 시간과 경험, 기억은 쌓이지 않고 휩쓸려 간다. 근린생활가로를 이루던 도시한옥 등 단층건물 대부분은 1900년대 초중반에 지어져 반세기도 채 버티지 못하고 건폐율 60%, 용적률 200%의 건물들에 자리를 내주었다. 그만큼 도시의 밀도는 높아졌고, 개량 목구조를 대신하여 적층이 가능한 현대적 재료와 구법들은 바삐 보급되었다.

　　그러나 단순히 법적 기준에 따라 쌓아 올린 건물들은 채광이나 환기, 조망 등 기본적인 내부 공간의 질적 수준을 갖추지 못했으며, 방치된 외부공간으로 인해 가로의 환경은 점차 더 악화되어갔다. 이러한 상황이 지속된 것은 경제논리와 수익의 문제로 귀결되는 대다수의 소규모 상업건물들이 상당 기간동안 특별한 경우를 제외하고는 건축의 영역으로 여겨지지 않았기 때문이다. 사회에 대한 건축의 윤리적 태도를 중시했던 경향은 건축이 임대업 등 부동산 자본 증식 도구로서 역할을 수행하는 것을 금기시해왔다. 대지의 일부를 공적 영역에 할애하는 방식의 작업들은 자본과 공공이라는 선악의 구도 안에서 건축의 공공성을 강조하였으나, 그와 같은 몇몇의 선례들도 상업 가로 문제의 보편적인 해법이 될 수는 없었다.

Preface What Remains of What Was There: The Case of the Commerical Streets

1989년 안암동 항공사진(아래), 2019년 안암동 항공사진(위) Courtesy of gis.seoul.go.kr

근린생활시설에서 다른 접근과 탐색이 활성화되기 시작한 것은 비교적 근래의 일이다. 새롭게 등장하는 세대의 건축가들은 기존의 업역에서 벗어나 설계와 개발-기획을 연계하는 등 건축의 외연 확장을 가속화하고 있다. 도시에서 사유화된 공간들이 차지하는 비중과 역할은 더욱 확대되었으며, 이들에게 있어 공과 사의 경계는 더 이상 이전처럼 명확하지 않아 보인다.

눈 여겨 볼 것은 그러한 변화 속에서도 건축의 유형들은 부분적으로 남아 지속되고 있다는 사실이다. 가로와의 관계, 외부공간의 활용과 향의 문제 등 리슈건축의 도시 상업가로 프로젝트들에서 나타나는 공통된 고민과 해법들 역시 지금까지 도시의 건축이 해결해야 했던 오래된 과제들과 다르지 않다. 다만 보다 더 복합적인 조건 아래에서 수행될 뿐이다.

후술할 세 개의 개념, 좌향과 여백 그리고 표층은 이처럼 변화된 상황 속에 놓인 리슈 건축의 전략으로서 여전히 작동하고 있는 유형의 조각들이다.

Just like the upstream of a river, where deposition does not take place due to its quick velocity, the time, experiences and memories of the place are not accumulated and carried away. Most of the single-story buildings, such as urban hanoks that constituted the neighborhood living street, were built in the early to mid-1900s and have gave up their spaces to buildings with a 60% building coverage ratio and 200% floor space index. Consequently, the density of the city increased, and modern methods as well as materials that can be stacked have been briskly distributed to substitute improved wooden structures. The following three keywords act as a common solution to these concerns and tasks, as well as the architectural language of Richue Architecture.

However, buildings that have been constructed according to legal standards have lacked the basic quality of interior spaces, such as lighting, ventilation or view, and the neglected external spaces have led to gradually deteriorated street environment. This situation continued because most small-scale commercial buildings that result in matters regarding economic logic and profit have not been considered as architectural domains except for special cases for a considerable period of time. The tendency to value the ethical attitude of architecture towards society has tabooed architecture to serve as a tool to increase real estate capital such as leasing. Works that have dedicated a part of their

안암동 「블랙박스 Black Box」, p.35

옥현동 「인터랙팅 큐브 Interacting Cube」, p.61

site to the public realm have emphasized the publicness of architecture within the good and evil structure of capital and public, but some precedents were unable to be a universal solution to the commercial street problem.

    Neighborhood living facilities have only begun to activate different approaches and explorations relatively recently. New generations of architects are accelerating the external extension of architecture by moving away from the existing areas and connecting design and development / planning. The proportion and roles taken up by privatized spaces in the city have been expanded further, and for them, the boundaries between public and private are no longer as distinct as before.

What we must pay attention to is the fact that the types of architecture partially remain and continue within such changes. The common problems and solutions that appear in the urban commercial street projects of Richue Architecture, such as the relationship with the street, the utilization of the external space and the matter of direction, are also not different from the old problems that urban architecture had to solve to this day. They are simply performed under more complex conditions.

    The following three keywords act as a common solution to these concerns and tasks, as well as the architectural language of Richue Architecture.

서문   남겨진 유형의 조각들: 상업가로의 경우

성수동 「더 그라운드 The Ground」, p.67

양재동 「보이드 라인 Void Line」, p.95

강릉 「레드 큐브 Red Cube」, p.103

전략 1

비틀어지는 도시의 좌향

도심 속에서 건축의 좌향坐向은 고정되어 있지 않다.
각층은 지면과의 거리에 따라 서로 다른 정면을 갖는다.
아래에서는 길과 사람을, 위에서는 빛과 풍경을 우선적으로 받아들인다.
위와 아래는 하나의 몸이지만, 허리를 비틀고 앉은 것처럼 서로 다른 곳을 바라보고 있다.
도시에서 건축을 경험하는 주체의 위치는 유동적으로 설정된다.

Strategies 1

Twisting and Turning: Reorientating Toward the City

The prospect of architecture is not fixed within a city.
Each floor gains a different front depending on the distance from the ground.
They first embrae the street and people from the bottom,
and the light and landscape from the top.
The top and bottom are one body, but they are each facing different directions as if they were sitting with their waists twisted.
The position of a subject experiencing architecture in the city is established flexibly.

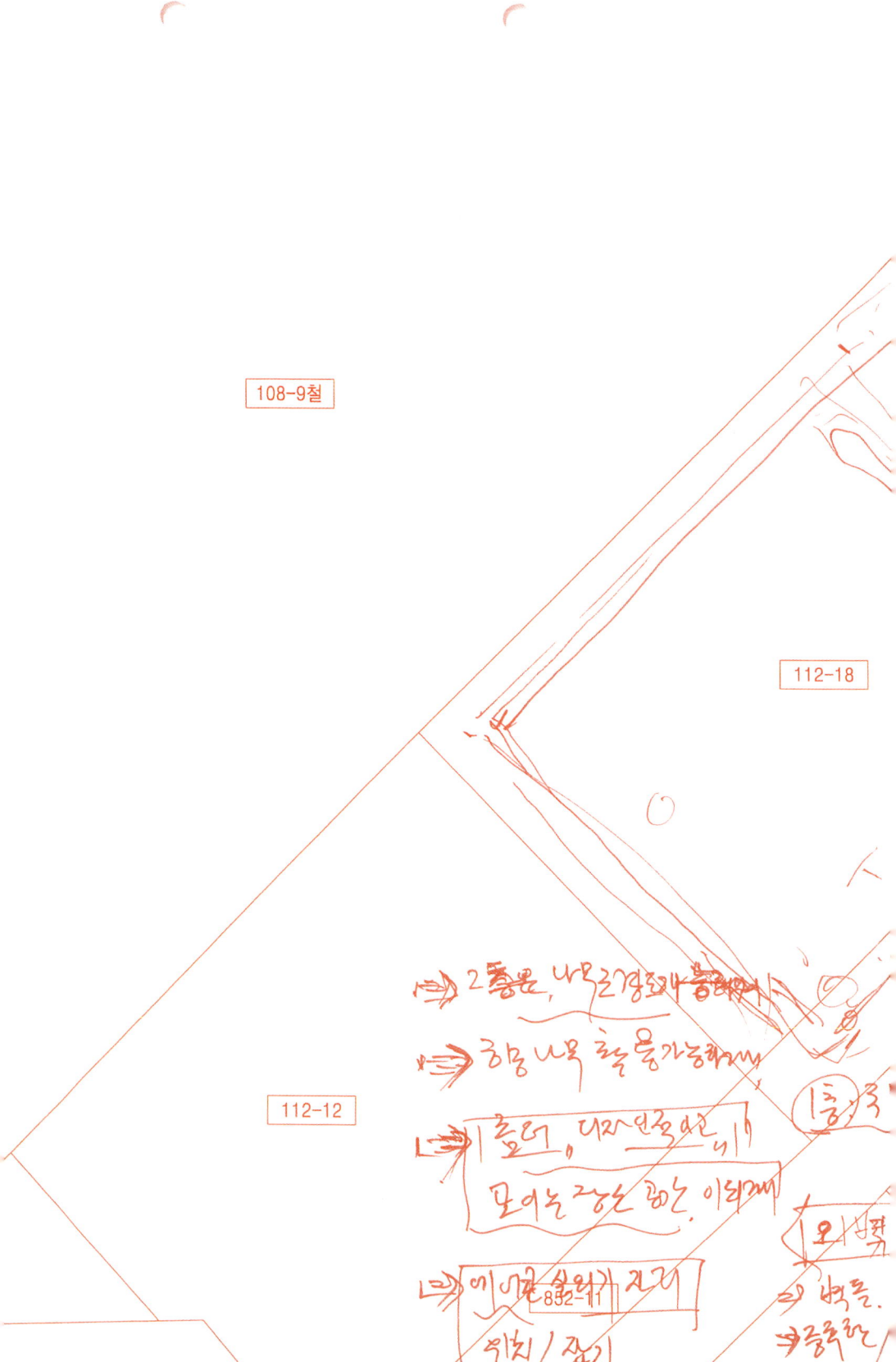

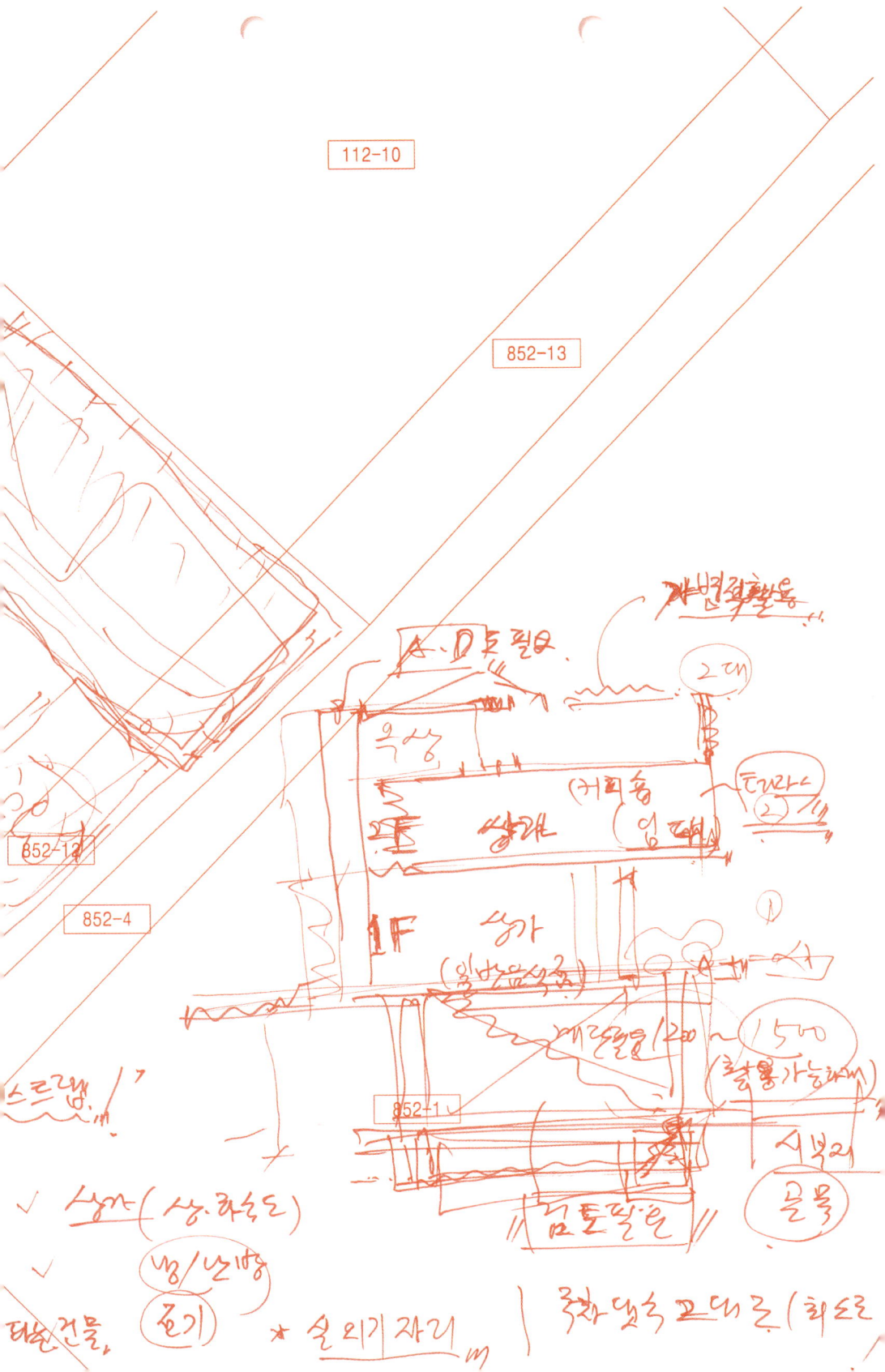

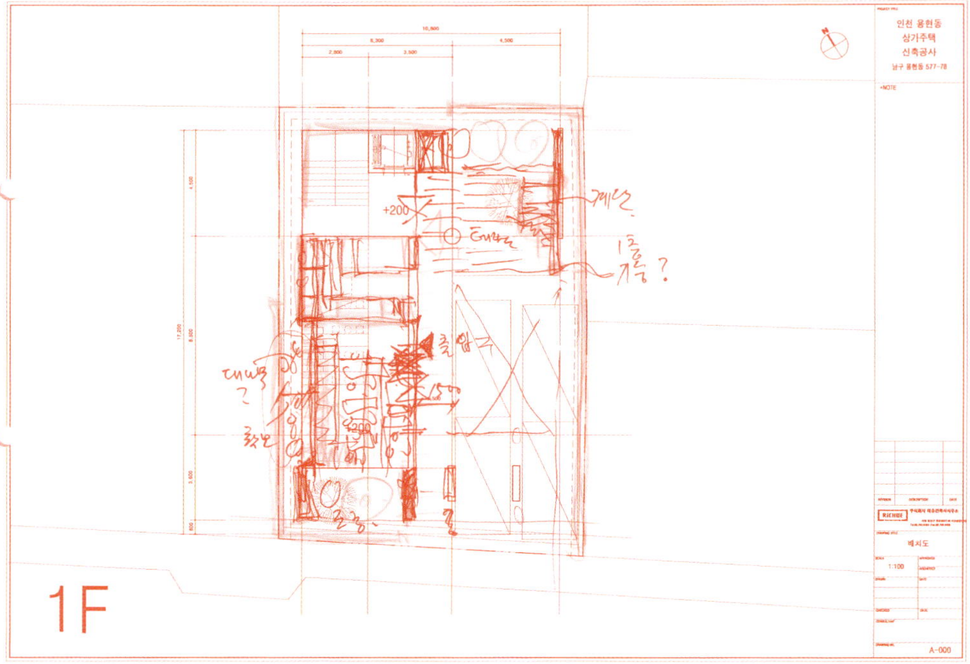

1F

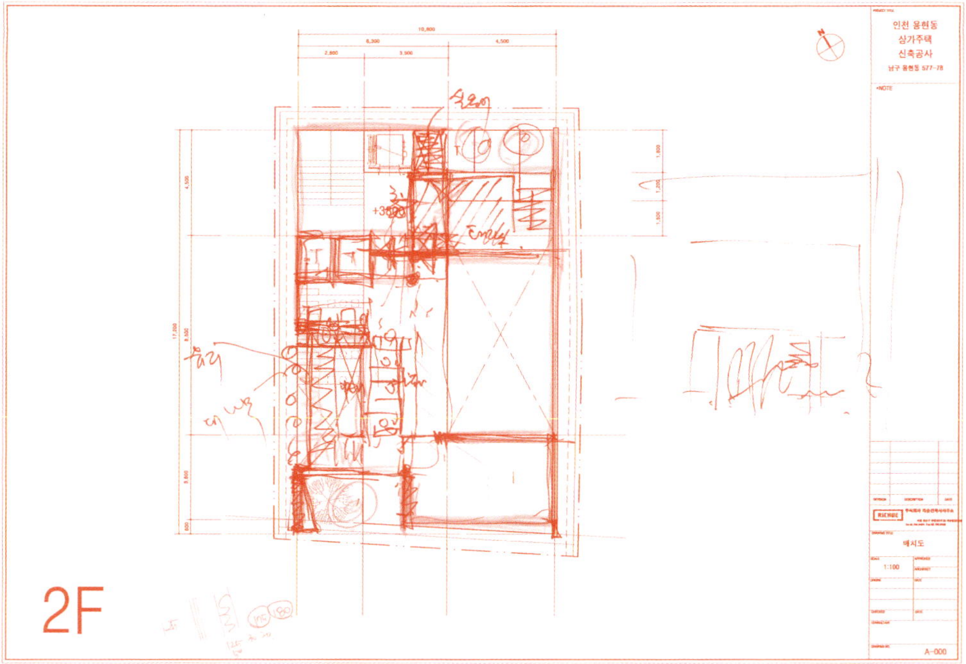

2F

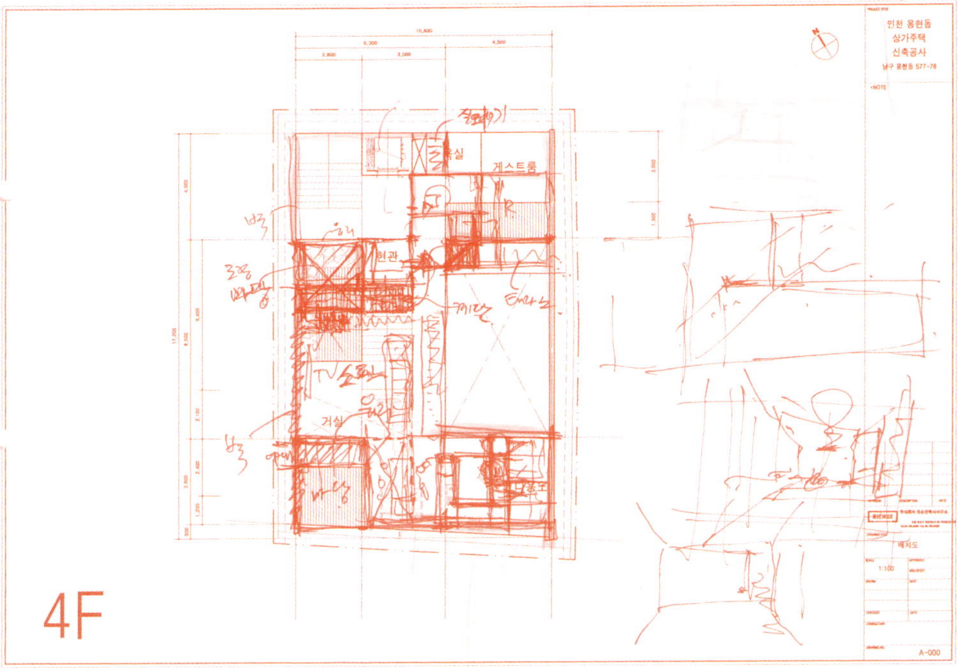

4F

5F

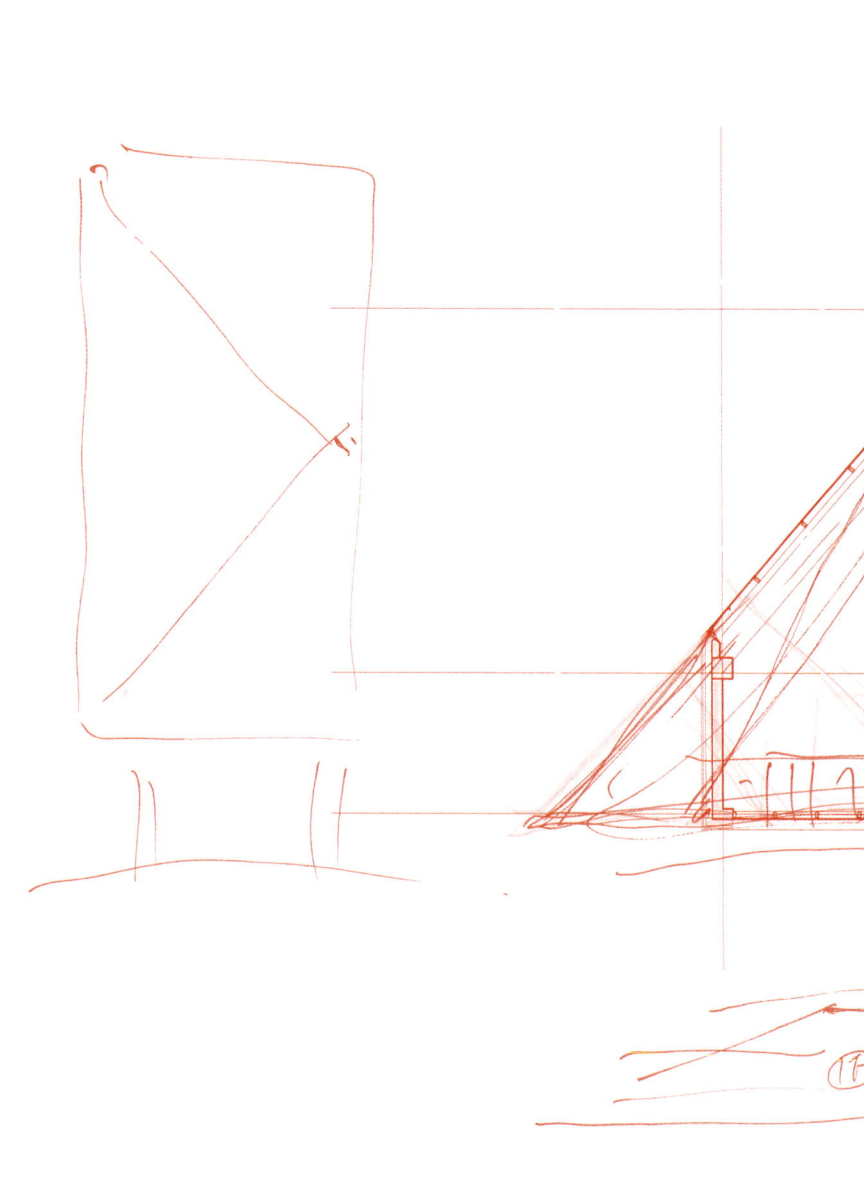

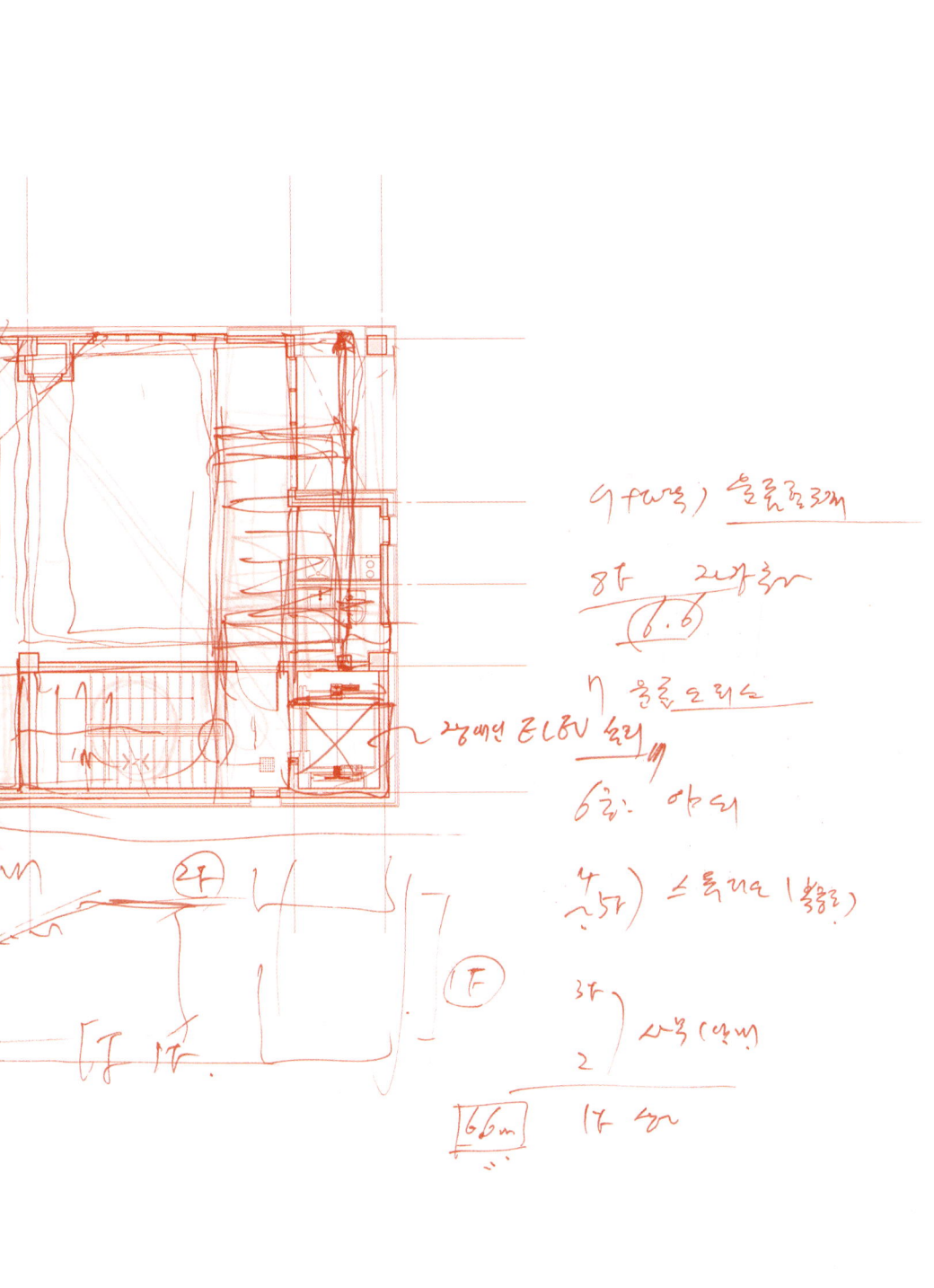

전략 1　　　　　　　　비틀어지는 도시의 좌향

안암동「블랙박스」의 남측 전경. 이웃한 한옥과 서로 다른 밀도를 갖고 있으나 외부 공간과 개구부의 방향과 그것을 구성하는 논리는 같다.

Southern view of Anam-dong 「Black Box」. Although it has a different density from the neighboring hanok, the direction of the openings and the exterior space, and the logic that forms them are the same.

닫혀진 열림

「블랙 박스」는 '한옥의 공간 구조 형식을 어떻게 조직할 것인가'라는 질문에서 시작했다. 작은 땅으로 인해 가운데 외부마당을 두기는 어려웠고, 구법상 목조가 아닌 3층의 철근 콘크리트조로 지어야 했다. 우리는 외부에 대해서는 닫혀진 형태와 내부로는 열린 공간의 관계를 조직하면서 공간을 구축해 갔다. 최대한의 건축면적으로 사각박스를 만들고, 검은 전벽돌로 외피를 둘러싸는 구법으로 바깥을 향해 닫혀 있는 미니멀한 형태를 만들었다. 그와 동시에 스킵플로어 방식으로 평면을 최대한 활용하며, 중심에 천창을 둔 비워진 공간으로 열린 내부를 만들 수 있었다.

Closed Openess

Black Box began with the question: "How should we organize the space, structure, and form of a Korean traditional house?" We could not afford to put a courtyard at the center of the small site, and had to take a feasible construction method by building a three-story reinforced concrete structure instead of a wooden structure. Organizing a closed form to the outside but an open spatial relationship to the inside, we proceeded with spatial construction. We created a rectangular box with its building coverage maximized and a minimal form clad with black traditional bricks so as to be closed to the outside. Meanwhile, we maximized the use of floor planning with split levels.

옥상에서 바라본 주변 건물들의 모습. 유리로 덮인 계단실을 남쪽에 배치하여 개구부로 활용한다. 그에 반해 대부분의 경우 창문 방향은 필지 형태와 상황에 의해 결정된 것에서 벗어나지 못하고 있다.

View of the surrounding buildings from the rooftop. The staircase covered with glass is placed to the south and used as an opening. On the other hand, the direction of the windows was unable to break away from being determined by the site's shape and situation.

테라스와 천창, 실외 복도를 통해 빛과 외기를 받아들인다.

Light and air are embraced through the terrace, skylight and external corridor.

우측에 위치한 한옥과 비교해보면, 외부공간의 배치 형식에 있어서 유사성을 지니고 있음을 확인할 수 있다.

f the plan of the exterior space's layout is compared to that of the hanok located side by side, one can see that they are facing the same direction.

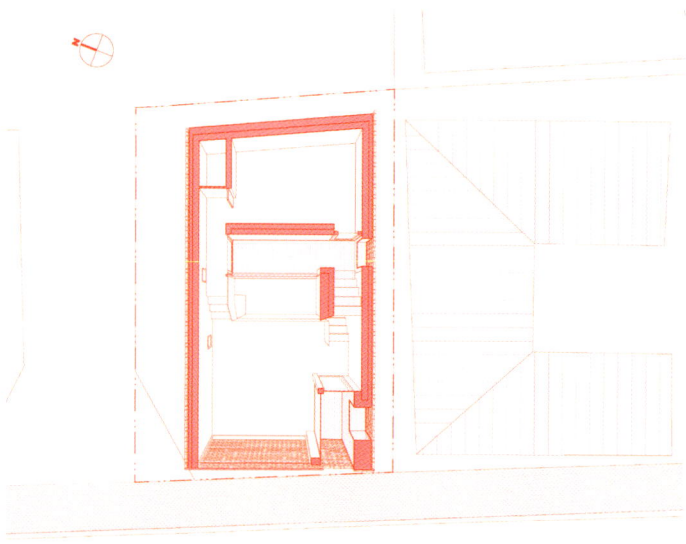

전략 1   비틀어지는 도시의 좌향

성수동 「더 그라운드」 진입부의 계단은 교차하는 2개의 길로부터 연속된다.

The stairs of the entrance to Seongsu-dong 「The Ground」 continue from two crossing roads.

파노라마 전경. 고층부의 향과 조망은 가로와 필지의 상황으로부터 독립적으로 설정된다.

Panoramic view. The direction and view of the upper floors are independently set from the situation of the street and the site.

수직 동선 다이어그램 스터디. 계단의 방향은 지면으로부터의 높이와 프로그램에 따라서 틀어진다.

Vertical circulation diagram study. The direction of the staircase changes depending on the height from the ground and the programs.

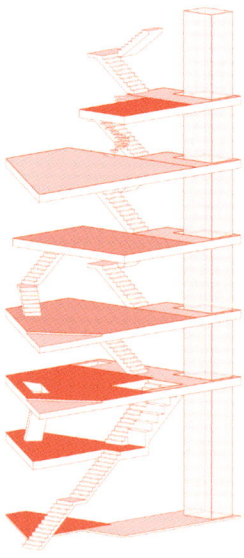

Strategies 1   Twisting and Turing: Reorientating Toward the City

도시와 관계조직 하는 큐브

「인터랙팅 큐브」의 계획은 전면도로가 좁고 긴 대지의 열악함에 대한 두가지 질문에서 시작되었다. 첫째, 어떻게 가로와 열린 관계를 조직할 것인가? 둘째, 3면이 인접필지로 둘러싸인 맥락에서 어떻게 풍부한 내외부 공간을 만들 수 있는가? 우리는 60%의 건축면적을 최대로 사용하면서 남은 40%의 비워지는 공간의 위치를 고민했고, 계획은 먼저 필지의 모양대로 큐브를 만든 뒤 건물의 중앙을 비워내는 순서로 진행되었다. 전체적인 형태는 가운데에 위아래를 관통하는 외부 공간을 두고 프로그램들 간의 시각적 관계를 고려하면서 가로를 향해 오픈되는 형식으로 만들어졌다. 비워진 공간들은 수직으로 적층된 각 프로그램의 사용주체들이 보이드와 도시의 풍경을 시각적-공간적으로 경험할 수 있도록 계획되었다. 이처럼 열린 경계를 가진 건물은 도시의 풍경과 보이드, 사용자와 보행자들의 관계를 조율하는 생활의 장치로서 작동하게 될 것이다.

A Cube That Organizes Relationships With the city

The plan of 「Interacting Cube」 began with two questions about the inadequateness of the site that is long and has a narrow front road. First, how will an open relationship with the street be organized? Second, how can ample interior and exterior spaces be created within a context where three sides are surrounded by adjoining lands? We thought about the location of the 40% of the emptied space that remained after using 60% of the building area as much as possible, and the plan was proceeded in the order of forming a cube in the shape of the site and emptying the center of the building. The overall form was made to be opened towards the street by placing an external space that penetrates the top and bottom in the center and considering the visual relationship between the programs. The emptied spaces were designed to allow the users of each vertically stacked program to visually and spatially experience the void and urban landscape. Buildings with such open boundaries will function as devices of life that coordinate the relationship between the urban landscape and voids, as well as the users and pedestrians.

전면의 가로에서 바라본 전경. 저층부의 상업 임대 공간은 길과 보행자를 향해 열려 있다.
Prnoramic view from the front street. The commercial rental spaces on the lower floors are open towards the road and pedestrains

전략 1　　　　　　　비틀어지는 도시의 좌향

상층부에 위치한 주거 영역의 정면을 바라본 모습.
View of the front side of the residential areas located on the upper floors.

5층의 거실은 건물 측면의 중정을 향하며 그로부터 자연광을 받아들인다.
The living room on the fifth floor faces the courtyard on the building's side and receives natural light from it.

용현동 「인터랙팅 큐브」의 등각투시도. 상부의 주거영역이 오픈된 방향이 저층부와 90도 틀어져 있다는 것을 확인할 수 있다.

Isometric drawing of Yonghyeon–dong 「Interacting Cube」. The direction in which the upper residential areas are opend is 90 degrees twisted from the lower floors.

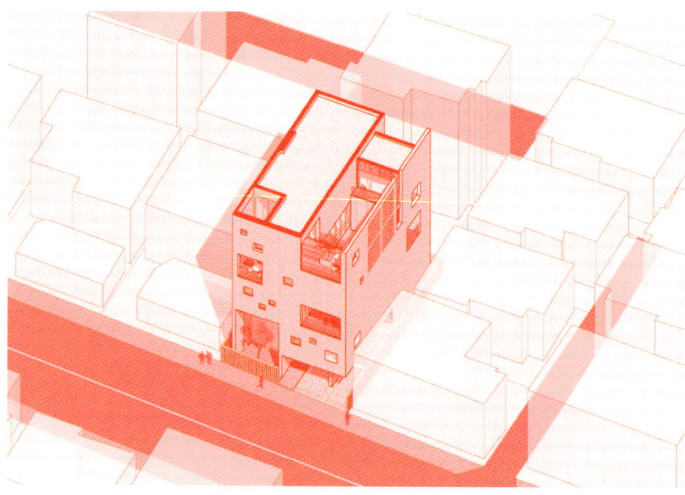

Strategies 1   Twisting and Turing: Reorientating Toward the City

풍경을 잇는 생성기하학

「레드 큐브」는 필지의 모양 그대로 용적을 채워 3개 층의 기하학적 형태가 이루었다. 이때 기하학은 행위를 규정짓는 추상적인 개념이 아니라 상업적 행위의 유동성과 변화를 담으면서, 지속적으로 도시풍경과의 관계를 조직한다. 대지 내에 느슨한 기하학적 장소를 생성하는 것이다. 필지 앞뒤에 면해 있는 두 개의 길은 도시의 풍경을 경험하는 데 있어 스케일에 차이를 갖고 있다. 1층은 뒷골목길의 풍경이 그대로 경험되는 방식으로 골목길 쪽으로 개방하였다. 넓은 전면 도로에서는 상대적으로 진입부를 거치는 반면 뒤편의 좁은 골목길에서는 넓게 개방된 공간을 만나게 되는 것이다. 한편 2, 3층은 강릉역 풍경과 주변의 도시풍경을 건물 안에서 바라볼 수 있다. 여기에서 창의 개념은 도시적 스케일로 확장된다. 건물 전체가 도시풍경을 내부로 담아내는 창인 동시에 큰 액자가 되어 강릉역에서 바라본 풍경 뿐 아니라 건물 이면 도시의 풍경까지 담아내는 큰 렌즈 같은 역할을 하는 것이다.

Generative Geometry Bridging Landscapes

「Red Cube」 was designed in a three-story geometrical form by maximizing the volume according to the shape of the site: here, the geometry is not an abstract notion that defines activities, but what keeps organizing the relationship with the cityscape while reflecting the flexibility and change of commercial activities. In other words, it is like a loose geometric place generated within the site. The two streets along the front and back of the site have different scales to experience the cityscape. The 1st (ground) floor was made open to the backstreet, so that you could experience the streetscape at the back as it is. While you should pass through the gateway along the broad frontal street, you can encounter a broadly open space along the narrow backstreet. Meanwhile, the 2nd and 3rd floors provide views from the inside to the landscape of Gangneung Station and the surrounding cityscape. Here, the concept of a window extends to the urban scale: like a window through which the cityscape comes inside, the whole building serves as a large lens that frames not only the landscape viewed from Gangneung Station but also the cityscape behind the building.

강릉 「레드큐브」 2층의 개방된 외부공간. 강릉역과 마주한 가로를 바라보고 있다.
View from the inside of Gangneng 「Red Cube」 to the front street.

강릉 「레드 큐브」의 단면 투시도. 상부의 임대 영역과 하부의 지층은 대지 앞뒤로 면한 넓은 대로와 좁은 골목길에 대하여 서로 다른 진입 방식을 설정하고 있다.

Section drawing of Gangneng 「Red Cube」. The upper rental floor and lower floor have different ways of entering from the front and back streets.

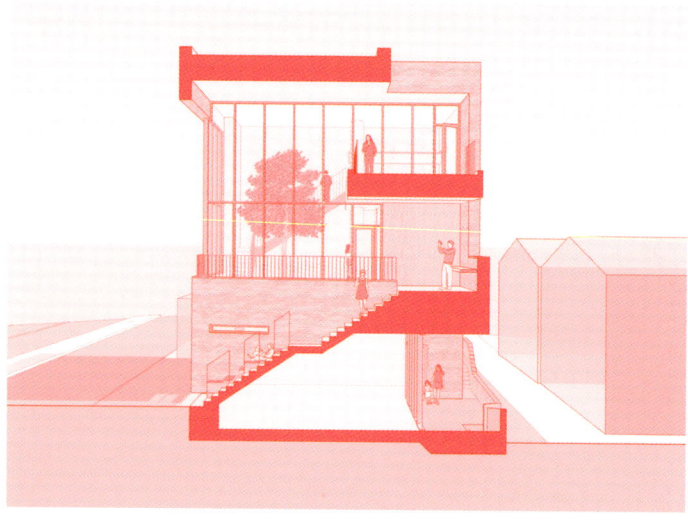

전략 1　　　　　　　비틀어지는 도시의 좌향

양재동 「보이드라인」의 북측의 가로에서 바라본 모습. 저층부 상업 임대 공간은 길과 보행자를 향해 열려있다.

View from the road on the north side of Yangjae-dong 「Void Line」. The commercial rental spaces on the lower floors are open towards the street and pedestrians.

남측의 자연채광과 열린 조망을 바라보는 5층 거실의 전경.
View of the living room on the fifth floor overlooking the natural light and open view from the south.

양재동 「보이드라인」의 단면도. 주거 영역인 상층부는 아래와는 반대로 남쪽을 향하고 있다.
Section drawing of Yangjae-dong 「Void Line」. The upper floors, which are residential areas, are facing the south in contrast to the lower floors.

Strategies 1   Twisting and Turing: Reorientating Toward the City

안암동 「블랙박스」

건축주는 옷을 디자인하고 샘플을 제작하며, 바이어와 상담하는 공간을 요구했다. 부부 중심으로 운영되는 작은 패션사옥으로, 일을 하면서 직원들과 같이 식사를 하고 휴식하는 공간이기를 원했다. 기존 공간이 한옥을 개조해서 운영해온 곳이라 마음에 들어 했고, 한옥처럼 밖에서는 패쇄적이고 안으로는 열린 공간이기를 바랐다. 23평의 작은 땅에서 여러 프로그램과 공간의 요구사항을 담아내야 했던, 간단치 않은 건축적 상상력이 필요한 프로젝트였다.

The client requested a space to design clothes, produce samples and consult with buyers. It was a small fashion office building run by a couple, and they wanted a space to work, eat and rest with the staff. They liked that the existing space was operated after renovating a hanok and they wanted the space to be closed from the outside and be an open space in the inside like a hanok. It was a complex project that required architectural imagination where we had to include the demands of various programs and spaces within a small land of 23 pyeongs.

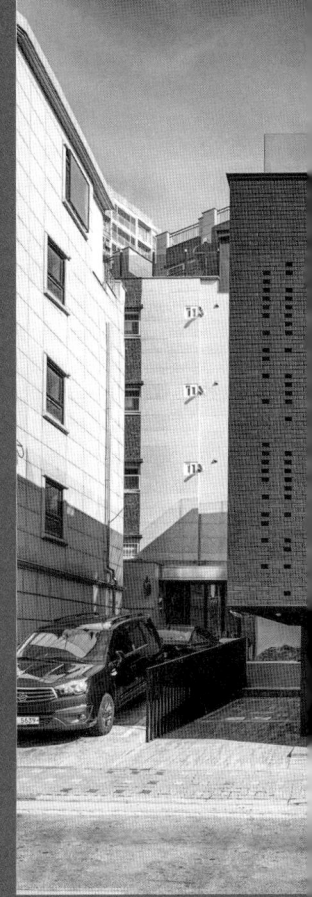

위치 서울특별시 성북구 안암동 **용도** 제2종근린생활시설(사무소) **대지면적** 77.40㎡ **건축면적** 44.41㎡ **연면적** 127.14㎡ **건폐율** 51.87% **용적률** 133.75% **규모** 지하 1층, 지상 3층 **구조** 철근콘크리트조 **시공** 본집

Location Goryeodae-ro 13-gil, Seongbuk-gu, Seoul **Use** Office **Site area** 77.40㎡ **Built area** 44.41㎡ **Total floor area** 127.14㎡ **Floor** B1F, 3F **Structure** Reinforced Concrete

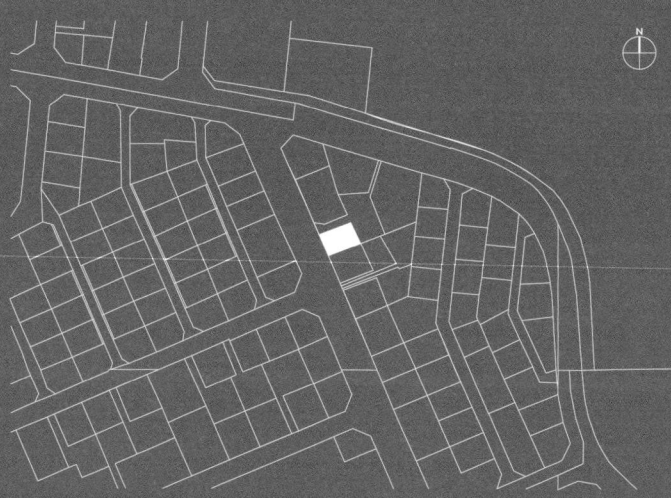

SITE PLAN

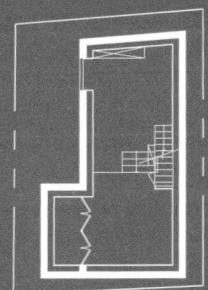
BASEMENT FLOOR PLAN

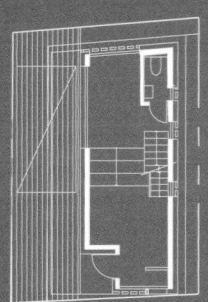
1st FLOOR PLAN

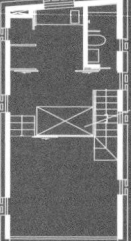

2nd FLOOR PLAN

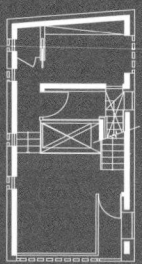

3rd FLOOR PLAN

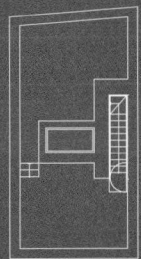

ROOF PLAN

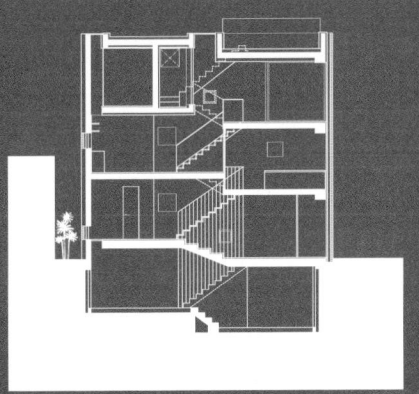

SECTION

전략 2

버려지지 않는 공동의 여백

건축으로 덮이지 않은 땅과 채워지지 못한 볼륨들은 모두의 것으로 남아,
끝내 누구의 것도 아닌 채로 버려진다.
문 밖에 내버려두지 않고 집 안으로 끌어들여 가둬질 때 비로소 건축의 여백은 빛과 바람이
통하고 누군가의 발길과 이야기가 머무는 자리로 쓰일 수 있다.
건폐율은 채우기 위한 법규regulation가 아니라 여백을 만드는 규율discipline이다.

Strategies 2

The Common Void: Space Beyond Properties

The lands and volumes that have not been filled with architecture remain as everyone's,
and are eventually discarded as nobody's.
Only when they are brought into the house rather than left outside the door can the empty
spaces of architecture be well lighted and ventilated, and used as a place where someone's
footsteps and stories remain.
The building coverage ratio is not a regulation, but a discipline to create the empty spaces.

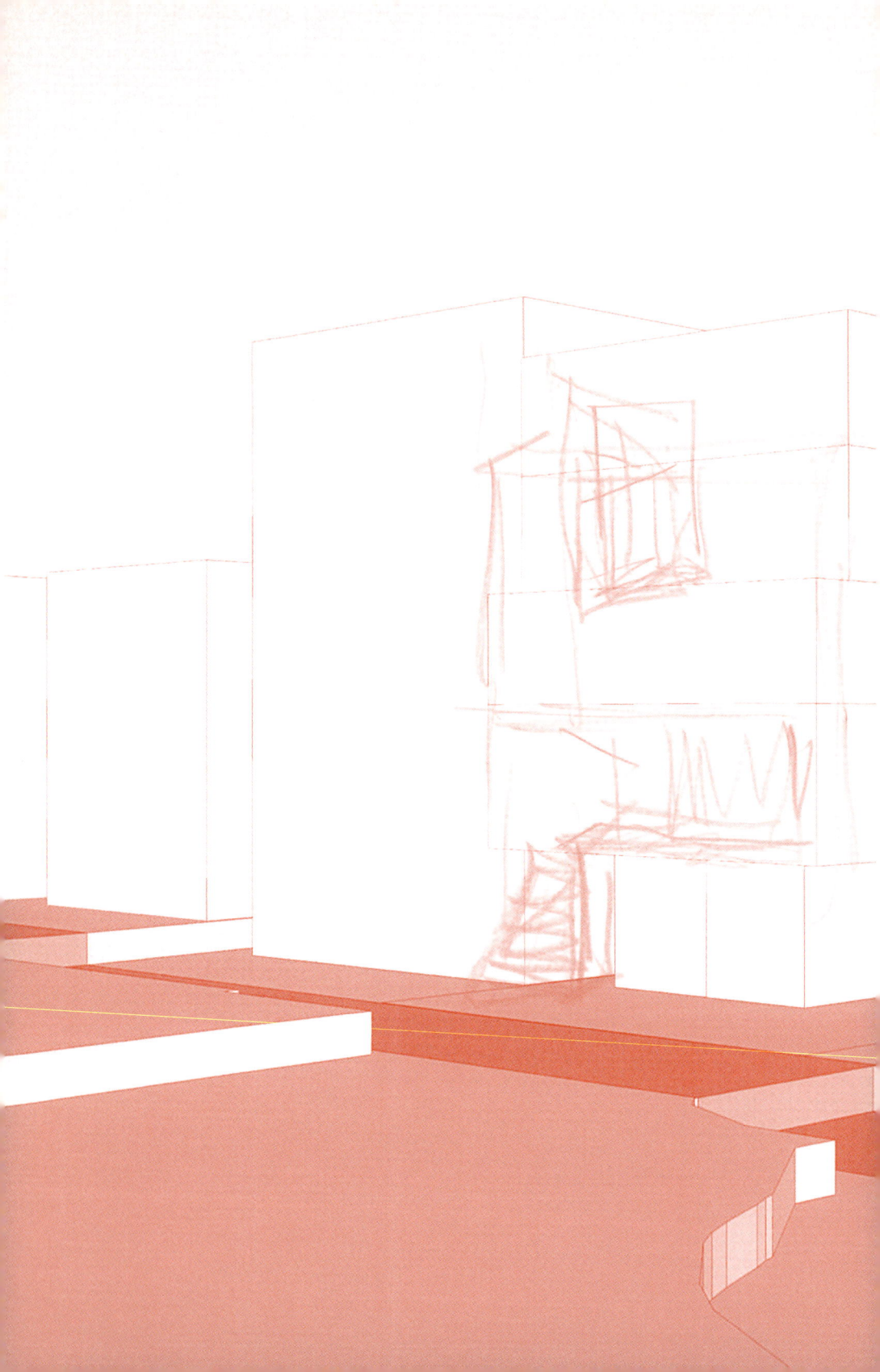

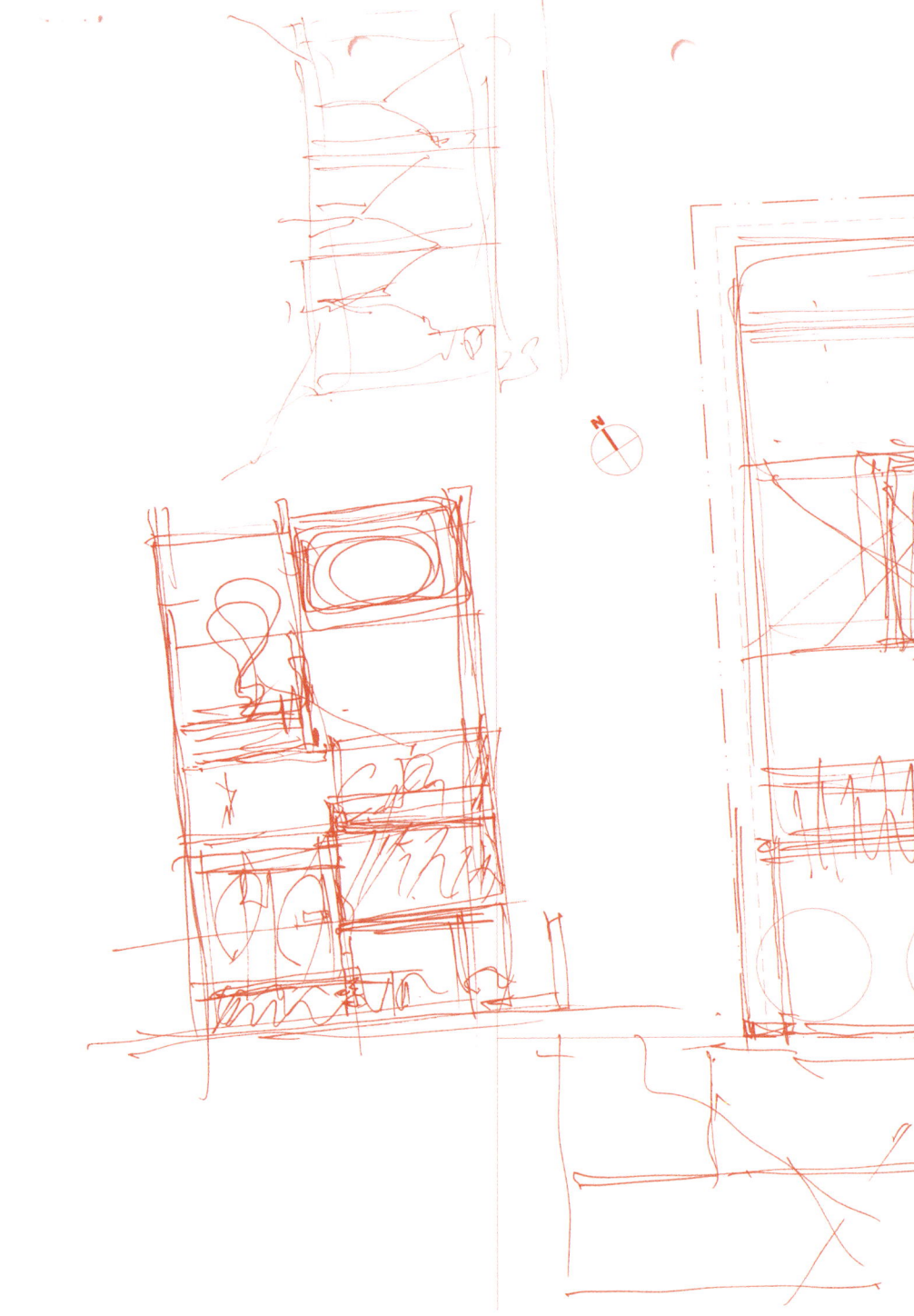

입체적인 소통의 공동적 보이드

긴 필지 위에 자리한 「인터랙팅 큐브」에서 중앙의 빈 공간은 수직적인 프로그램들 간의 입체적인 소통의 공간으로 기능한다. 내부의 공간들은 수직의 보이드에서 바로 연결되어 풍부한 빛의 경험과 시각적 확장성을 경험하게 된다. 층별 테라스 사이의 입체적인 소통은 다른 프로그램 간의 관계를 조직하면서 보이드를 중심으로 커뮤니티를 형성하게 된다. 전면의 가로는 시각적으로 연계되어 도시의 풍경을 입체적이고 깊이감 있게 담아낸다. 이처럼 입체적인 소통을 가능케하는 수직의 보이드는 가로를 향해 열린 경계와 상호 작용하면서 공간적, 시각적 경험과 풍경을 만들고 있다.

Common Void of Three-Dimensional Communication

The empty space in the center of 「Interacting Cube」, which is placed on a long site, functions as a space of three- dimensional communication between vertical programs. The interior spaces are directly connected in the vertical voids, allowing the ample light and visual expansion to be experienced. The three-dimensional communication between the terraces on each floor will organize relationships between other programs and form a community centered on the void. The street at the front is visually connected, and this three-dimensionally captures the urban landscape with depth. The vertical void that allows three-dimensional communication interacts with the boundaries open towards the street and creates spatial and visual experiences and landscapes.

「용현동 인터랙팅큐브」를 수직으로 관통하는 중정은 각층의 프로그램에 따라 서로 다른 기능으로 작동한다.
The courtyard that vertically penetrates 「Yonghyeon-dong Interacting Cube」 serves as a different function depending on the programs of each floor.

지면에서 중정을 올려다본 모습. 외부공간을 둘러싼 4개의 면이 다양한 방식으로 열리고 닫힌다.

View of the courtyard from the ground below. The four sides that surround the external space open and close in various ways.

외부공간들은 중정에 의해 연결되어 서로 소통이 이루어질 수 있도록 계획되었다.

The external spaces on each floor are designed to connect and communicate with each other with the courtyard.

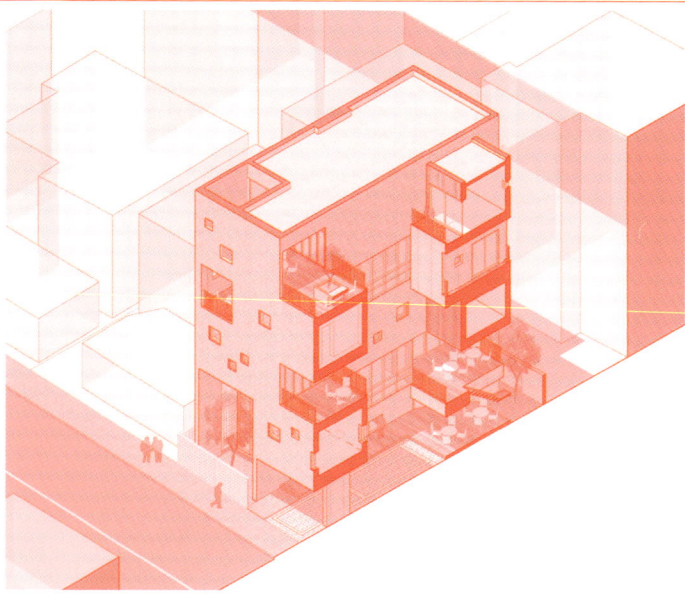

Strategies 2   The Commom Void: Space Beyond Properties

생성적 장소인 수직 보이드

「더 그라운드」는 세 개의 보이드 공간을 두고 있다. 이들 각각은 프로그램과 주변 환경의 관계를 조율하면서 정체성을 갖는다. 저층부(1~3층)의 주차장과 계단실은 전시 프로그램과 도시가로와의 관계를 맺어주고 있다. 주차장의 보이드는 경계를 투명한 유리로 계획하여 건물을 둘러싼 세 면의 도시가로까지 시야를 열어준다. 이 곳에서 이루어지는 프로그램이 주변의 길에서도 투영되는 것이다. 가로와 이어지는 계단실의 보이드는 교육과 전시 영상이 가능한 스탠드형으로 구성하였으며, 그 아래로는 주차장으로 연결되어 입체적인 장소성을 갖는다. 중층부(4~5층) 내부의 보이드는 촬영 및 업무 프로그램이 자연채광과 갖게 되는 관계를 고려했다. 이 장소는 높은 층고와 자연광을 받는 벽면 등 비일상적이고 독특한 공간감을 만들어 주고 있다. 끝으로 상층부 (6~7층)의 테라스는 이벤트와 휴식, 도시풍경의 조망 등 도시적 스케일의 관계를 고려하여 계획했다. 이 공간은 주변의 저층 건물 높이를 반영하여 조화를 이루면서 3면을 유리로 둘러 한강, 남산, 아차산 등 서울의 도시 풍경을 개방감 있게 담아내는 장소가 되었다. 이처럼 건축면적이 작고 수직적인 성격이 강한 건물에서 비어있는 영역들은 프로그램과 관계된 생성적 공간이 되어 도시를 표상한다.

Vertical Void, a Generative Place

「THE GROUND」 has three void spaces. Each of these spaces gain identities as they coordinate the relationship between the program and surrounding environment. The parking lot and staircase of the lower floors (first to third floors) are establishing a relationship between the exhibition program with the urban streets. The void of the parking lot was planned with transparent glass to open the view to the three urban streets surrounding the three sides of the building. The program that takes place here is projected onto the streets around it. The void of the staircase that leads to the street consists of a stand type that allows education and exhibition projects, and it gains a three-dimensional locality as it is connected to the parking lot below. The void inside the middle floors (fourth to fifth floors) considered the relationship that the filming and work programs have with natural light. The studio area creates an unusual and unique sense of space, with its high floor height and walls that receive natural light. Finally, the terrace on the upper floors (sixth to seventh floors) was planned while considering its relationship between the urban scale, such as events, relaxation and the view of the urban landscape. As this space reflected the height of the low-rise buildings in its surrounding, harmonized with them, and had three of its sides surrounded by glass, it became a place that openly captures the urban landscape of Seoul, such as

사선으로 도려낸 외부공간을 통해
들어오는 도시의 풍경

The cityscape entering through the external space that was diagonally removed.

the Hangang river, Namsan and Achasan mountains. In buildings like this with small architectural areas and strong vertical personalities, the empty areas become a generative space related to the programs and represent the city.

길 건너편에서 바라본 성수동「더 그라운드」의 야간 전경. 법적 기준에 포함되지 않는 건축의 외부 공간들은 하부의 주차공간과 상부의 테라스로 계획되었다.

Night view of Seongsu-dong「The Ground」from the other side of the street. The external spaces of the building, which are not included in the legal standards, were planned as the parking space of the lower part and the terrace of the upper part.

Strategies 2　　　　The Commom Void: Space Beyond Properties

길과 이어지는 공동의 보이드

「레드 큐브」에서 건물로 접근하는 길은 계단으로 이어진다. 여기서 계단은 중요한 공동적 요소로 기능하고 있다. 계단에서부터 3층까지 비워진 공간에서 건물의 사용자와 주변의 풍경은 서로 오버랩되며, 외부 공간은 건물 안으로 깊숙이 들어와 안으로 이어진다. 직접적인 상행위가 없더라도 누구나 건물의 안이자 밖인 테라스에 들어올 수 있는 것이다. 반면, 3층에는 내부를 통해서만 밖으로 나갈 수 있는 테라스가 있고 안팎의 외부공간들은 비워진 건축의 영역에 의해 연결된다. 이 공간들은 사유화된 건물 내에 있음에도 익명의 사람들과 공유되는 공동의 여백으로 도시가로에 새로운 경험을 주는 장치가 되고 있다

Common Void That Connects With the Path

In the 「Red Cube」, the path to the building is led by a staircase. Here, the stairs function as an important common element. In the empty space from the stairs to the third floor, the building users and the surrounding environment are overlapped with each other, and the external space enters deep into the building and leads to the inside. Even if there is no direct commercial activity, anyone can enter the terrace that is the inside as well as the outside of the building. On the other hand, there is a terrace that can only be entered through the inside on the third floor, and the external spaces of the inside and outside are connected due to the area of the emptied architecture. Although these spaces are in a privatized building, they have become a device that provides new experiences to the urban streets with a common empty space that is shared with anonymous people.

전면 가로에서 바라본 강릉 「레드큐브」의 진입부. 건물의 앞뒤로 관통하는 외부공간은 공적 영역과 사적 영역이 중첩하는 공동의 여백이다.

View of the entrance of Gangneung 「Red Cube」 from the front street. The exterior space that penetrates the front and back of the building is a common blank space where the public and private areas overlap with each other.

전략 2   버려지지 않는 공동의 여백

양쪽으로 열린 3층의 테라스는 건물이 가로막고 있는 전후면의 도시의 풍경을 이어주고 있다.

The terrace on the third floor that is open on both sides connect the landscape of the city of the front and back sides, which are obstructed by the building.

내부와 외부는 투명한 유리면 하나만을 사이에 두고 연속된다.

The interior and exterior continue with a single transparent glass side in between.

입체적 소통의 수직 보이드

「보이드 라인」의 흐트러진 형태와 적층된 매스들은 계단을 비롯한 동선과 중정에 의해 관통되며 수직적인 관계를 형성한다. 길에서 연속되는 계단은 보행자를 2층으로 끌어들이며 상가의 접근성을 높인다. 동시에 2층의 사용자와 가로의 사람들 그리고 3층의 시선들은 서로 교차되며 수직적으로 소통하게 된다. 4, 5층에 비워진 공간은 남매 세대에게 자연광을 제공하는 중정이자 내부적인 소통의 공간이다. 이 중정은 채광이나 환기뿐 아니라 내부의 계단과도 접해 있어 다양한 시각적인 경험을 거주자에게 제공할 것이다. 수직적 보이드는 이처럼 프로그램 간의 시각적, 공간적 소통 장치로서 사용자들은 이를 통해 공동의 경험을 만들어 나간다.

Vertical Void of Three-Dimensional Communication

The disordered forms and stacked masses of 「Void Line」 form a vertical relationship as they are penetrated due to circulations including the stairs and the courtyard. The stairs that continue from the street are attracting pedestrians to the second floor and are increasing accessibility to the shop complex. At the same time, the eyes of the second-floor users, the people on the street and the third floor cross and meet each other and communicate vertically. The emptied space on the fourth and fifth floors is a courtyard as well as a space of internal communication that provides natural light to the siblings' households. This courtyard not only functions for lighting and ventilation but will also provide the residents with various visual experiences since it adjoins the interior staircase. Through the vertical void, as a visual and spatial communication device between programs, users can create a common experience through this.

전면 가로와 접해있는 양재동 「보이드 라인」의 테라스. 수직으로 열린 공간이 연속되며 시야를 열어준다.
The terrace of Yangjae-dong 「Void Line」 adjoining the front street. The vertically open space continues and opens the view.

중정은 또한 2개 층의 주거 영역이 서로 소통하는 영역이기도 하다.

The courtyard is also an area where the two-story residential area communicates with each other.

두 개의 단면도는 저층부와 상층부의 서로 다른 외부 공간의 활용 방식을 보여준다.

The two section drawings show the different ways each of the lower floors and upper floors use the external space.

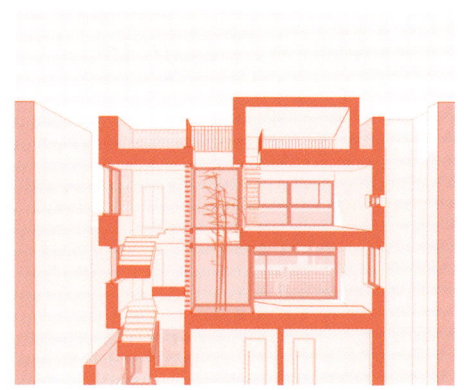
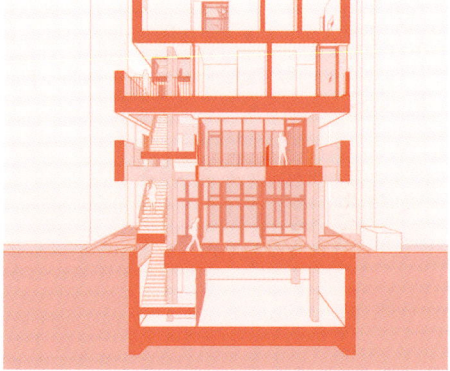

용현동 「인터랙팅 큐브」

인천시 남구 준주거 지역에 위치한 땅으로, 15미터 전면도로에 접한 좁고 긴 직사각형 필지이다. 1, 2층의 카페, 3, 4층의 근린생활시설, 5층의 주택이 건축주로부터 요구된 프로그램이었다. 전면도로에 접해 있는 폭이 좁고 안쪽으로 긴 형태의 필지가 가진 열악함을 건축적으로 해결해 주기를 요구했다. 양 옆과 뒤쪽에 인접 필지가 있어 미래에 건물이 들어서면 필지 주위가 패쇄적으로 둘러싸일 것을 걱정했다.

The site is located in a semi-residential area in Nam-gu, Incheon, and it is a long and narrow rectangular site. The programs requested by the client were a cafe on the first and second floors, neighborhood living facilities on the third and fourth floors, and a house on the fifth floor. They requested that we architecturally solve the inadequateness of the narrow site, which adjoins the front road, that is shaped long inward. Since there are adjacent lands on either side as well as the back, we were worried that the area around the site would seem closed if a building was built in the future.

위치 인천광역시 미추홀구 용현동 **용도** 제2종근린생활시설, 단독주택(다가구) **대지면적** 240.40㎡ **건축면적** 143.21㎡ **연면적** 567.74㎡ **건폐율** 59.57% **용적률** 236.00% **규모** 지상 5층 **구조** 철근콘크리트조 **시공** (주)예지인 종합건설

Location Michuhol-gu, Incheon **Use** Restaurant, House **Site area** 240.40㎡ **Built area** 143.21㎡ **Total floor area** 564.74㎡ **Floor** 5F **Structure** Reinforced Concreate

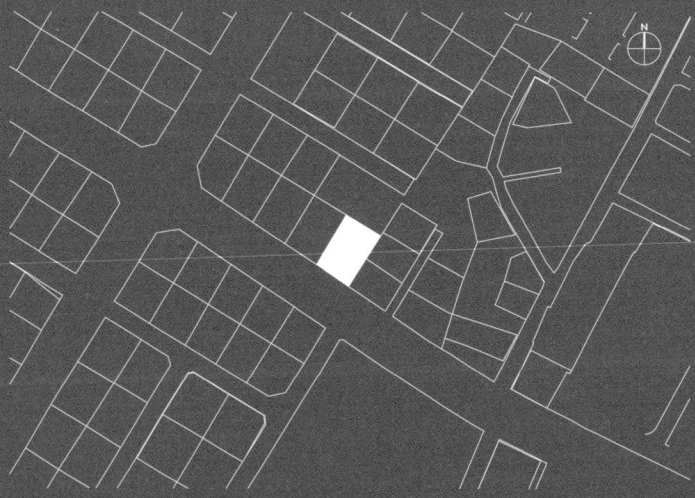

SITE PLAN

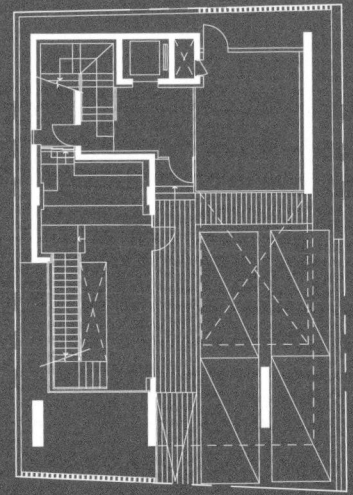
1st FLOOR PLAN

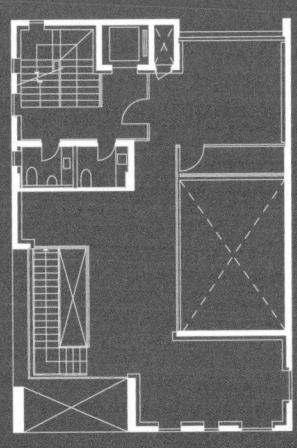
2nd FLOOR PLAN

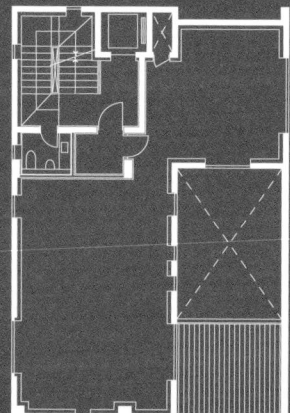

3rd FLOOR PLAN

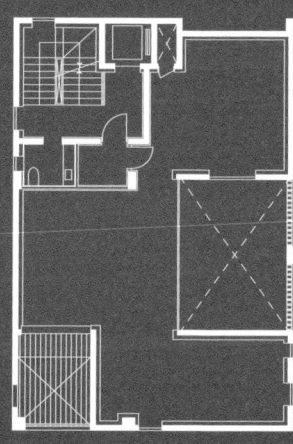

4th FLOOR PLAN

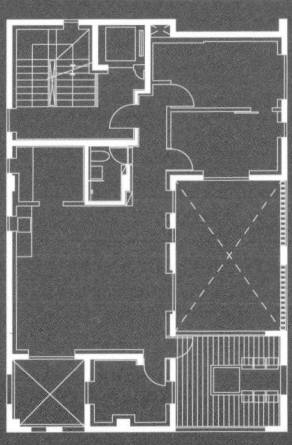

5nd FLOOR PLAN

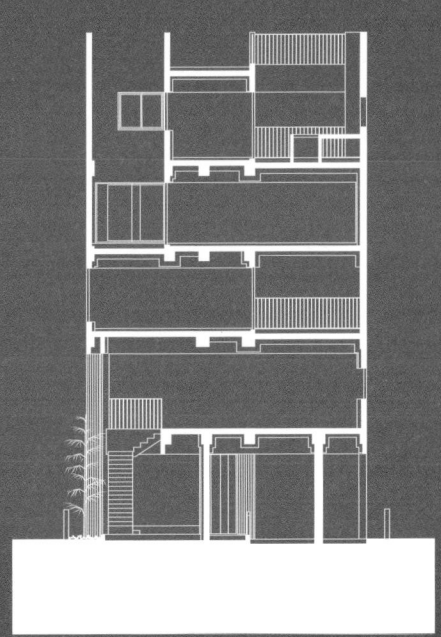

SECTION

성수동 The Ground

근래 주목받고 있는 장소인 성수동. 건축주는 뉴욕과 한국을 오가면서 작업하는 사진작가이다. 그는 임대, 전시, 작업실, 거주, 파티룸 등 다양한 프로그램이 담긴 건물을 원했다. 풍부한 문화적 감수성이 있는 성수동의 특징을 담는 공간을 만들고 싶어 했다. 작은 땅에서 높이 솟아오른 수직적인 이 건물은 건축주가 창간한 잡지 『THE GROUND』로부터 설계 개념의 모티브를 얻었다.

Seongsu-dong, a place that is attracting attention recently. The client is a photographer who works and travels between New York and Korea. He wanted a building with various programs, including rentals, an exhibition, workshop, residence and party room. He wanted to create a space that contains the characteristics of Seongsudong, which has a rich cultural sensitivity. The design motif of this tall and vertical building, which rises from the small land, was inspired by 『THE GROUND』, a magazine published by the client.

위치 서울특별시 성동구 성수동2가 **용도** 제2종근린생활시설, 단독주택(다가구) **대지면적** 190.18㎡ **건축면적** 113.53㎡ **연면적** 845.47㎡ **건폐율** 59.70% **용적률** 392.13% **규모** 지하 1층, 지상 9층 **구조** 철근콘크리트조 **시공** ㈜예지인 종합건설

Location Seongsui-ro, Seongdong-gu, Seoul **Use** Neighborhood facility, Multi-household house
Site area 190.18㎡ **Built area** 113.53㎡ **Total floor area** 845.47㎡ **Floor** B1F, 5F
Structure Reinforced Concrete

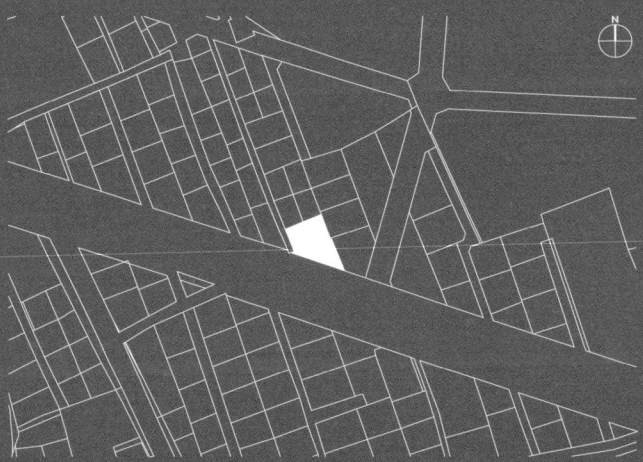

SITE PLAN

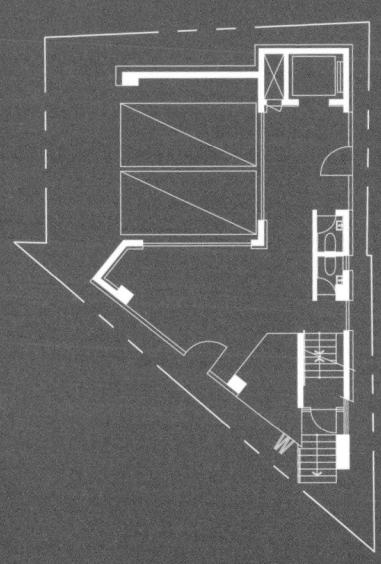
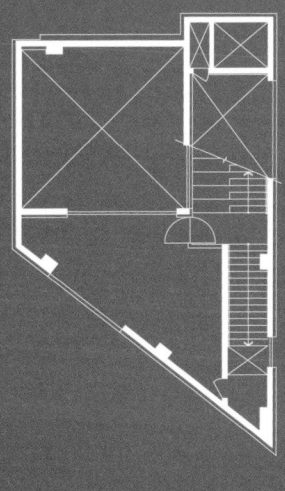

1st FLOOR PLAN **2nd FLOOR PLAN**

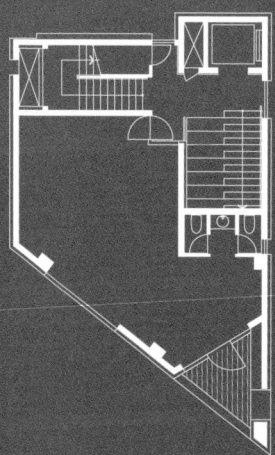
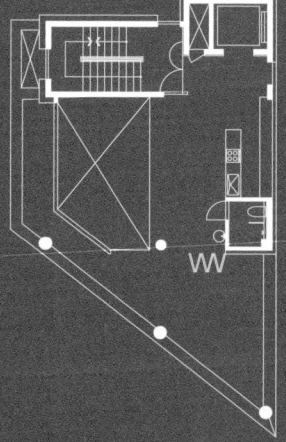

3rd FLOOR PLAN

6th FLOOR PLAN

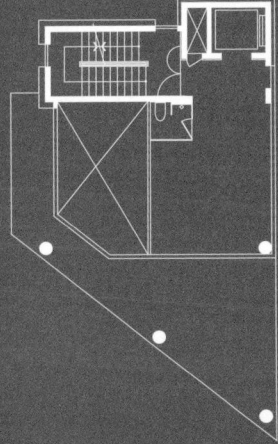
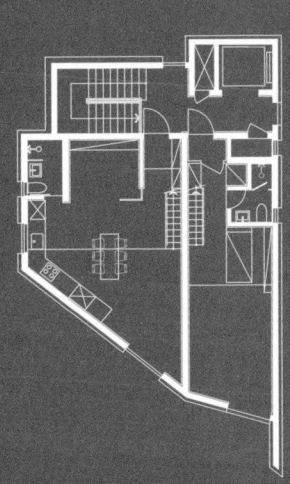

7th FLOOR PLAN

8th FLOOR PLAN

SECTION

전략 3

소비되는 표층의 두터움

여러 겹으로 이루어진 건축은 다채로운 경계를 지니고,
그 사이에 위치한 공간들은 안과 밖을 오가며 둘의 거리감을 중재한다.
유리 벽 뒤편 깊은 곳에서 비친 몸짓들이 두터운 표층을 넘어 거리의 시선을 이끄는 사이,
안에서 바라본 가로의 풍경은 건축의 경계들 위로 포개어진다.
경계는 이제 안과 밖을 구분하기보다 그 사이의 매개체interface로서 기능한다.

Strategies 3

Consumed Yet Tangible: Thick Surfaces

Architectures formed in multiple layers have various boundaries,
the spaces located between them mediate the distance between the inside and the outside.
While the gestures reflected from deep behind the glass wall draw the attention of the street beyond the thick surface, the streetscape seen from the inside is overlapped onto the boundaries of the architecture.
The boundaries function as the interface between the inside and the outside, rather than distinguish between both.

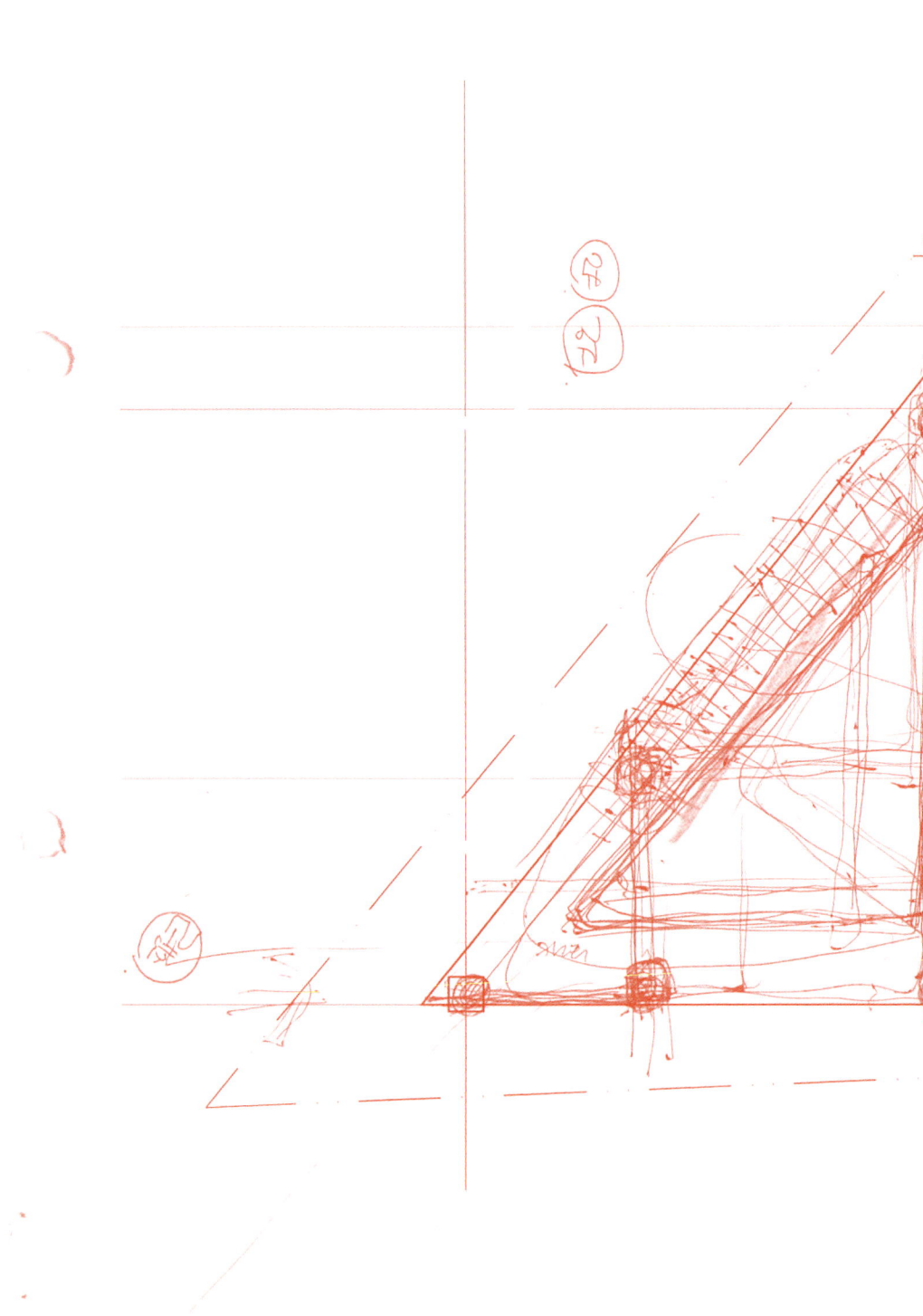

스프링클러 설치기준 (6장 자동차 등 자재)
누수이든 2개동 이상 제4권 205

가짜개 내부에도.. 설치.

15인승
(장애인용)

나

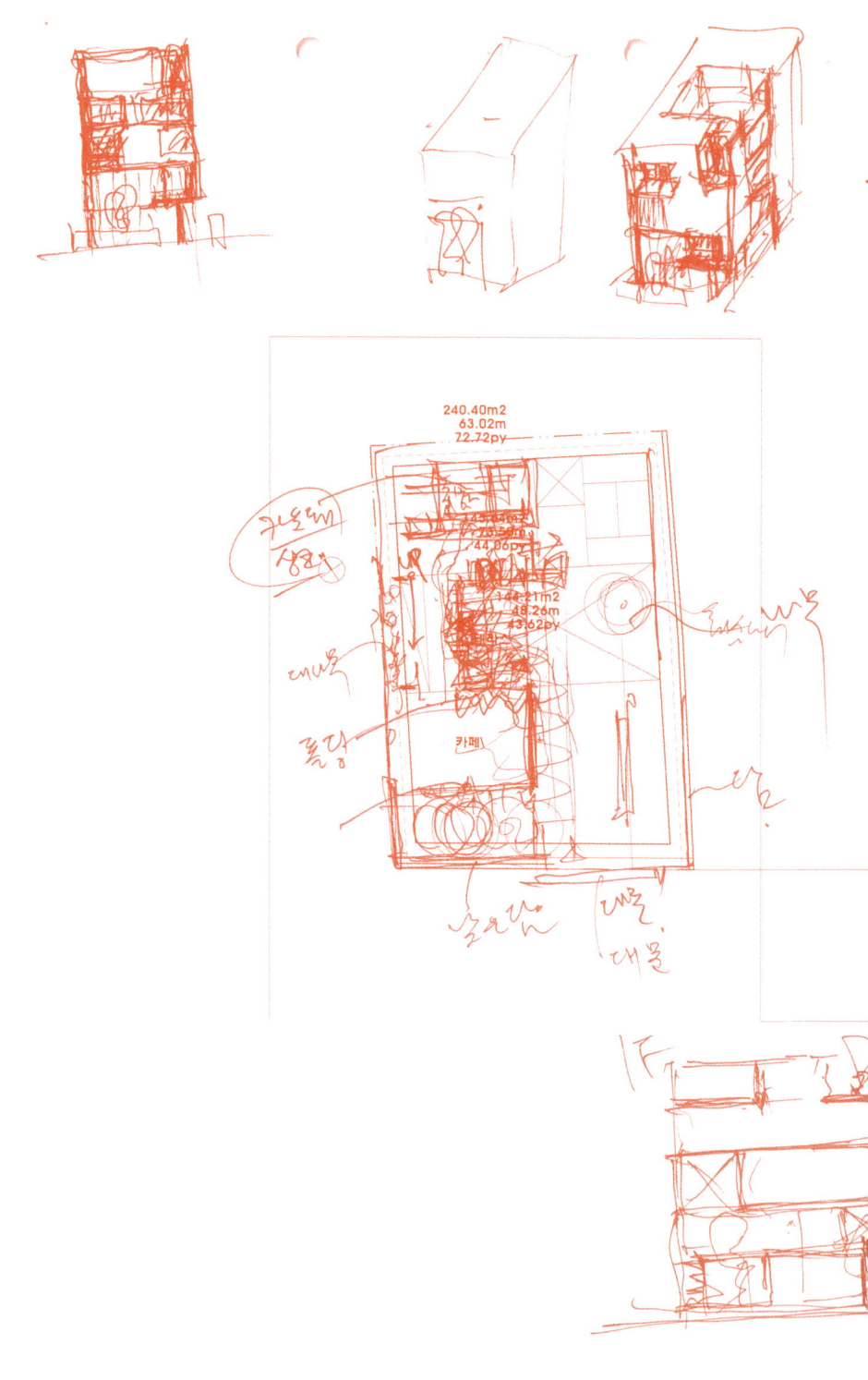

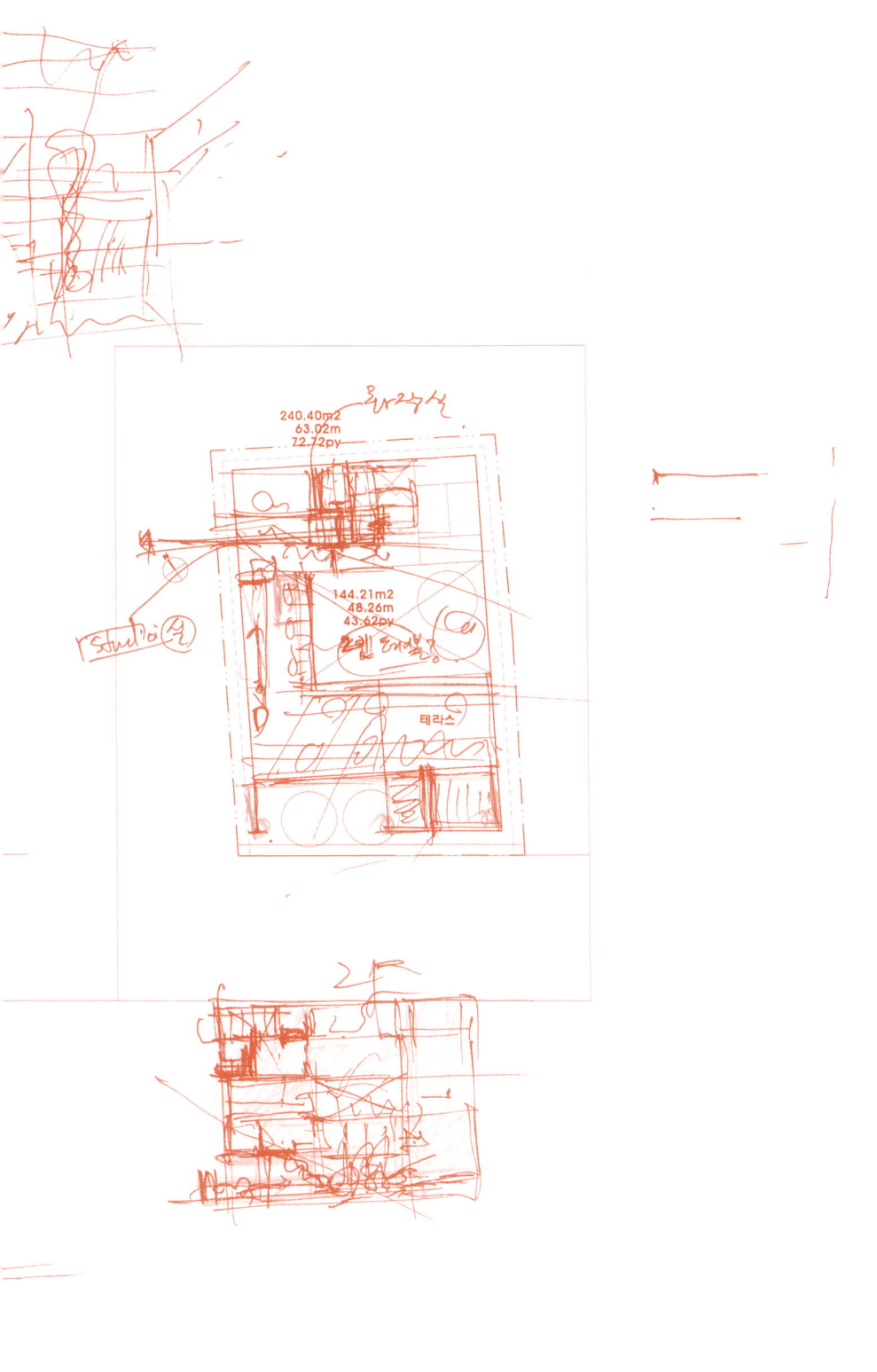

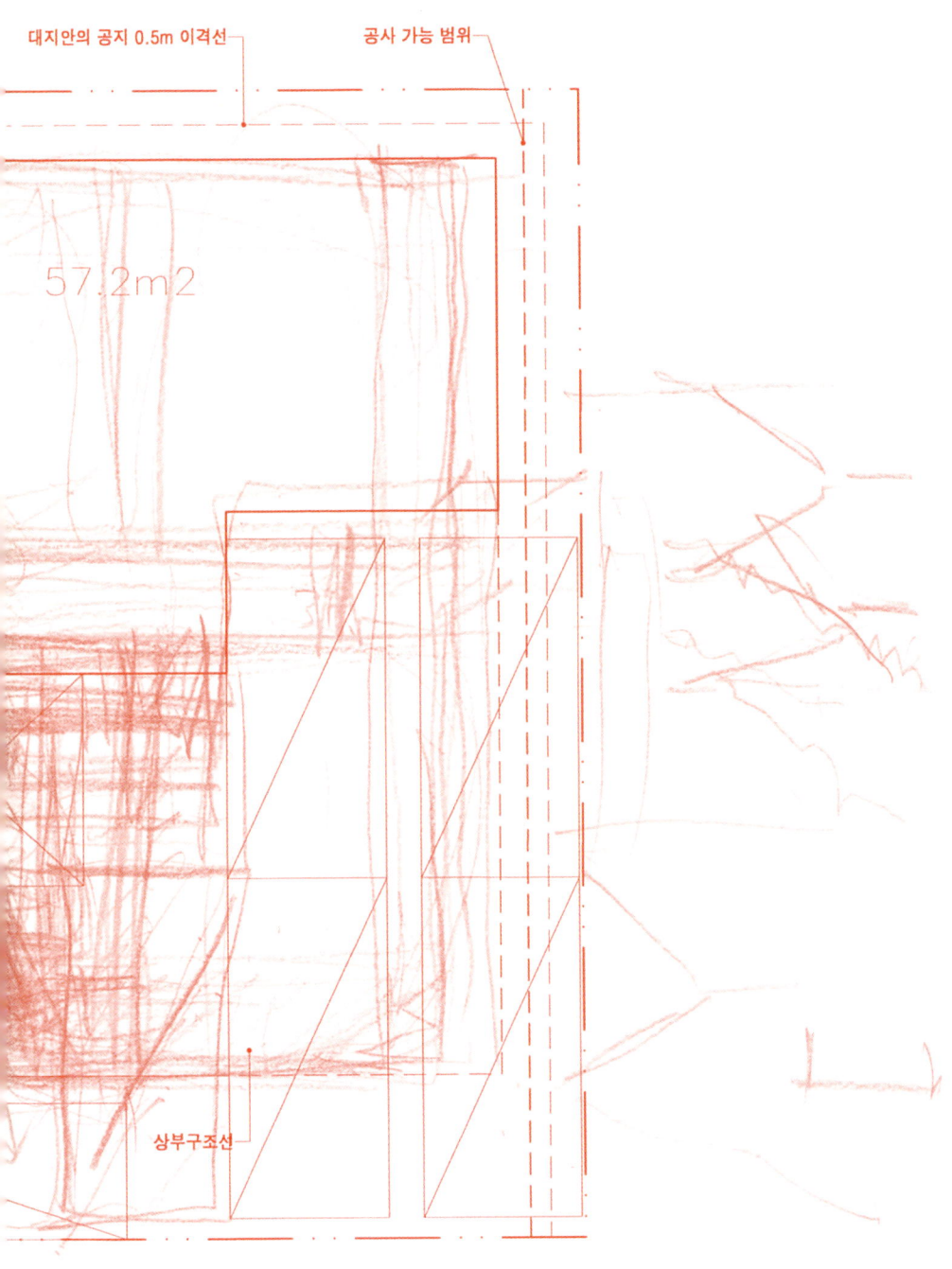

도시 필지 간의 유연한 경계

수평적으로 구성된 「보이드 라인」은 다양한 깊이감을 가진 경계들의 밴드로 구성되어 있다. 표층의 개념은 도시 필지들 사이의 경계와 관계에 대한 새로운 건축적 제안이다. 경계의 밴드는 롱브릭의 가로줄눈만을 강조하여 쌓은 매스의 선들과 대조되며 수평적인 깊이감을 더하고 있다. 매스의 선을 이루는 구멍난 블록들은 빛의 산란이나 시각적 중첩 등을 일으키며 공간의 경험을 다채롭게 한다. 「보이드 라인」은 버려지기 쉬운 도시 필지 간의 경계를 시각적, 공간적 경험이 이루어지는 유연한 관계의 장으로 활용하고 있다.

Flexible Boundaries Between the Urban Sites

The horizontally formed 「Void Line」 is composed of bands of boundaries with various depths. The concept of surface is a new architectural proposal for the boundaries and relationships between urban sites. The bands of the boundaries contrast with the lines of the masses stacked while emphasizing the horizontal joints of the long bricks and adds horizontal sense of depth. The perforated blocks that form the lines of the masses cause scattering of light, visual overlap, etc., and diversify the experiences of the space. 「Void Line」 is using the boundaries between urban sites that are inclined to be abandoned as a place of flexible relationships where visual and spatial experiences take place.

「양재동 보이드라인」은 가로를 향해 하나의 면을 이루고 있는 주변의 건물들과 달리 여러 층이 중첩되어 있다.
Yangjae-dong 「Void Line」 has multiple overlapping floors, unlike the buildings in its surrounding that have one side towards the street.

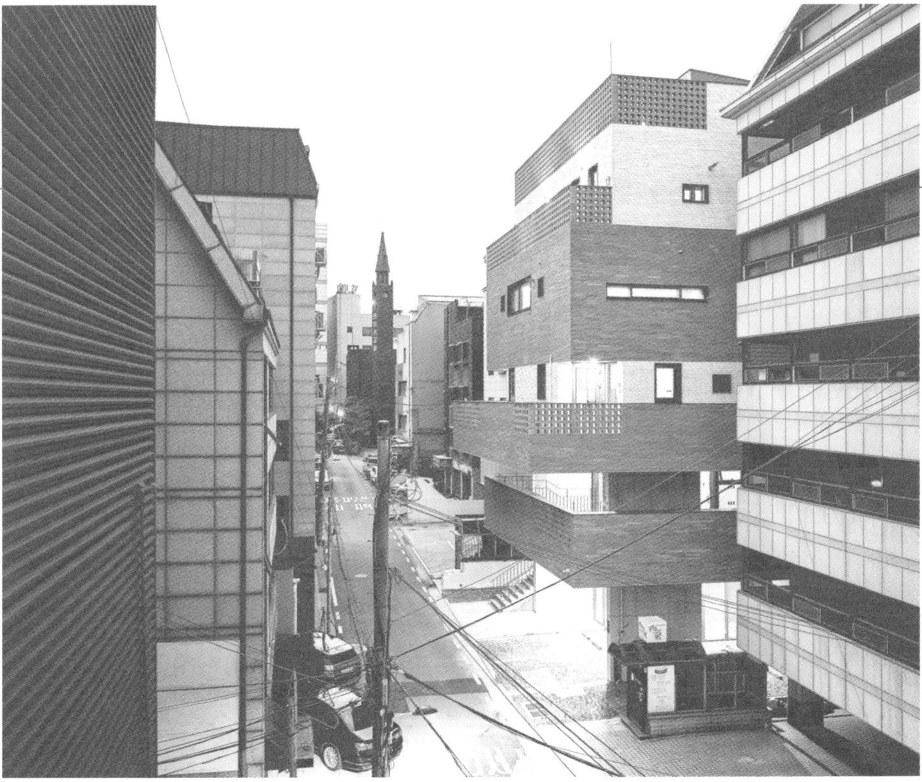

Strategies 3 　　　Consumed Yet Tangible: Thick Surfaces

양재동 「보이드라인」이 갖는 안과 밖의 두터운 경계 공간
Thick boundary space of the interior and exterior of Yangjae-dong 「Void Line」.

투어리즘적 경계 장소

강릉역에서 나오면 곧바로 눈에 들어오는 「레드 큐브」의 필지는 현지인들을 상대로 상행위를 하기 보다는 여행객들을 위한 장소가 되는 곳이다. 새로운 도시를 만날 때 역에서 느끼는 여행의 기분과 설레임, 기다림의 공간을 만들고자 했다. 건물에 채워지는 상업의 종류보다는 공간성 자체로 드러나는 투어리즘적 장소를 고민했다. 강릉의 풍경을 담으면서 편안한 기다림의 장소를 만들기 위해 건물의 안과 밖은 경계 지어지지 않은 모호한 영역으로 만들어졌으며, 투명한 유리에 의해서만 안과 밖을 구분하며 내외부의 중첩된 공간 경험을 제공하고 있다. 건물은 현지인과 여행객, 사용자와 익명인, 안과 밖의 사람들이 자연스럽게 섞이며 중첩되는 모호한 경계의 경험을 제공하며 투어리즘의 장소로 만들어진다.

Touristic Boundary Place

The site of 「RED CUBE」, which is seen as soon as one exists Gangneung station, is a place for tourists rather than a place that engages in commercial activities with locals. We wanted to create a space with the feeling of travel, excitement and waiting felt at the station when meeting a new city. We thought about a touristic location that is revealed through the spatiality itself rather than the types of commerce that fills the building. In order to create a comfortable waiting place that includes that landscapes of Gangneung, the interior and exterior of the building were made into boundless and ambiguous areas, and as the interior and exterior are separated only with the transparent glass, an accumulated spatial experience of the inside and outside is provided. The building provides an experience of ambiguous boundaries where locals and tourists, users and anonymous people, as well as people inside and outside naturally mix and overlap, creating a touristic place.

전면 가로에서 바라본 모습

View from the front street

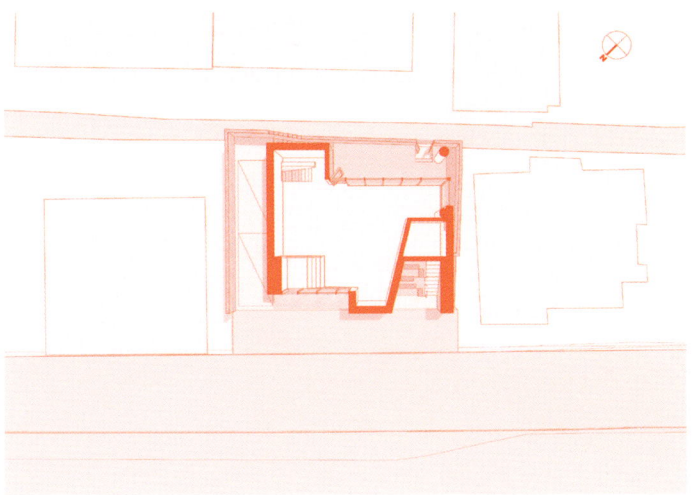

강릉 「레드큐브」의 1층 평면 투시도

First floor perspective plan of Gangneung 「Red Cube」

담장과 테라스, 계단 등은 건물을 둘러싸고 있는 조건에 따라 서로 다른 표층을 형성하고 있다.
Above. Walls, terraces, stairs, etc. each form different surfaces depending on the conditions surrounding the building.

Strategies 3　　　　　Consumed Yet Tangible: Thick Surfaces

디지털 그라운드 매트릭스 외피

「더 그라운드」는 세 면이 길로 둘러싸인 작은 땅에 비해 상대적으로 높게 지어지는 건물이므로, 어느 방향에서도 시야에 잘 들어오는 조건을 가지고 있다. 형태는 필지의 모양을 따랐지만, 다른 방식을 통해 사진작가의 사옥이 갖는 의미를 담아내고 싶었다. 우리는 건물의 외피가 디지털 이미지를 이루는 화상의 최소 단위인 픽셀의 이미지로 보여지도록 함으로써, 프로그램과 그 안에서의 행위를 깊이감 있게 전달하는 입면을 계획하였다. 건물의 외관은 채워진 영역과 비워낸 공간이 수직적으로 혼재되어 있다. 주차장을 활용하는 1, 2층의 전시장과 4,5층에 위치한 복층의 촬영장 그리고 6, 7층의 오픈된 이벤트 장소는 솔리드한 픽셀이미지를 나타내는 외피와 함께 표층을 구성하고 있다. 디지털 그라운드라는 개념으로 구성된 입면은 불규칙한 모양을 가진 타일 모듈이 빛의 각도에 따라 다르게 연출하는 표면을 배경으로 다양한 행위가 중첩되는 깊이감을 드러낸다.

Digital Ground Matrix Surface

「THE GROUND」 is a building that was built relatively high compared to the small land, which has three of its sides surrounded by streets, and so it is easily seen from any direction. Although the form followed the shape of the site, we wanted to contain the meaning of a photographer's office building through different methods. By making the exterior of the building in the image of pixels, the smallest unit of pictures that form a digital image, we planned a facade that delivers the programs and activities inside in depth. The filled areas and emptied spaces are vertically mixed in the exterior of the building. The exhibition hall on the first and second floors that utilizes the parking lot, the split-level studio on the fourth and fifth floors and the open event venue on the sixth and seventh floors form the surface with the building's exterior that represents a solid pixel image. The facade, which was composed with the concept of a digital ground, reveals a sense of depth where various actions are overlapped with the background of the surfaces that are portrayed differently by the irregular tile modules depending on the angle of the light.

투명하고 열린 공간과 외벽의 픽셀 이미지의 병치는 내부의 행위와 프로그램을 밖으로 내비치는 수단으로 이용되었다.

The juxtaposition of the transparent and open space and the pixel image of the external wall was used as a way to indicate the internal actions and programs to the exterior.

평면 투시도. 여러 겹으로 둘러싸인 공간의 구조를 보여준다.

Perspective plan. The structure of the multi-layered space is shown.

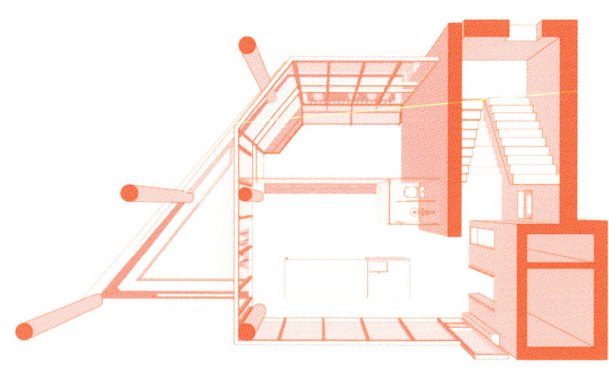

불규칙하게 돌출된 외피와 뒤편의 길의 모습은 건물 전면의 깊이감을 준다.

The irregularly projecting envelope and the appearance of the street at the back provides a sense of depth to the building's front.

생성적 건축 표층

「안암동 블랙박스」의 미니멀하고 심플한 형태와 창호 계획은 유리와 벽돌 외피가 중첩된 더블스킨의 표층을 만들고 있다. 전벽돌 텍스쳐에 페인트 도장으로 균질한 벽돌의 색감과 질감을 만들었다. 특히 도로에 면해 중첩되어 있는 벽돌 면은 자연스럽게 흐르는 패턴을 가진 다공 쌓기를 통해 닫혀 있는 도시가로를 향해 낮과 밤이 다른 표층을 연출한다. 균질한 물성으로 구축된 건축 표층은 햇빛과 조명을 걸러내며 내부공간과 외부가로에 대한 생성적 역할을 감당하게 된다.

Generative Architectural Surface

The minimal and simple plan for the doors and windows of 「Anam-dong Black Box」 are creating a double-skinned surface where the glass and brick external wall is overlapped. The traditional brick texture was painted to create a homogeneous brick color and texture. In particular, the overlapping brick wall that faces the road portrays different surfaces by day and night towards the closed city streets through the porous stacking that has a naturally flowing pattern. The architectural surface that is built with homogeneous properties filters the sunlight and lighting and plays a generative role for the interior space and exterior streets.

안암동 「블랙박스」의 야간 전경. 더블스킨의 벽돌 패턴은 낮과 밤의 다른 가로의 모습을 연출한다.
Night view of Anam-dong 「Black Box」. The double-skin brick pattern creates appearances of the street that are different by day and night.

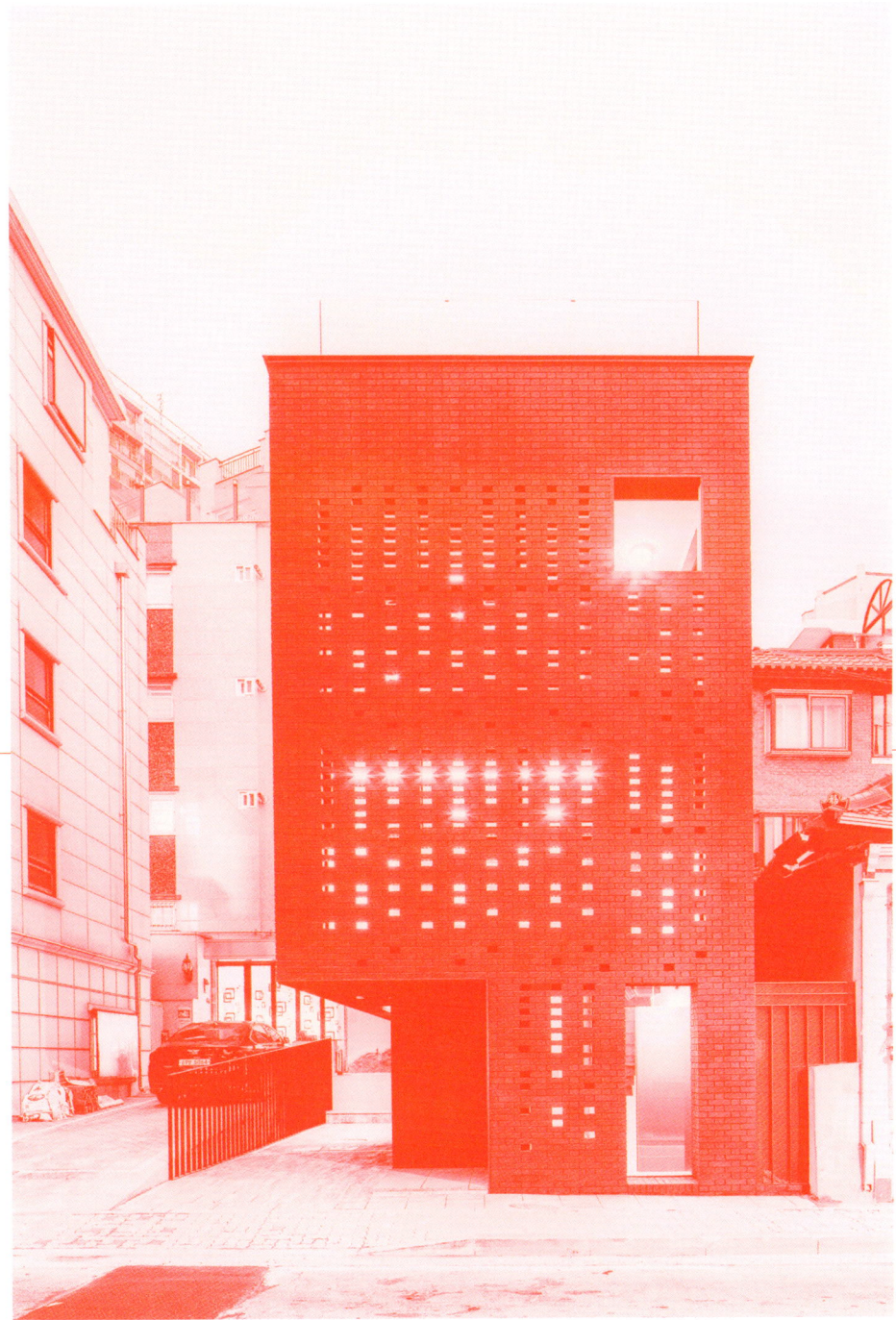

다공의 벽돌벽은 바깥의 시야를 차단하는 동시에 빛을 받아들이는 기능을 한다.
The porous brick wall serves the function of receiving light while blocking the view from the outside.

전략 3   소비되는 표층의 두터움

양재동 Void Line

북측이 전면도로인 대지이고, 저층부 상권이 잘 형성된 이면
가로에 접해 있다. 건축주인 남매는 상층부 4, 5층에 각각
거주하고자 했다. 지하와 1, 2층은 근린생활로 3층은 임대
원룸으로 구성하기를 원했다. 디자인이 잘 되어 임대
시장에서 경쟁력이 있고, 그와 동시에 거주환경이 좋은
건물이기를 원했다.

The site has a front road to the north, and it adjoins a side
street with a well-developed low-rise business district.
The clients, who are brother and sister, wanted to each live
on the upper fourth and fifth floors. They wanted to
organize neighborhood living facilities on the basement,
first and second floors, and rental studios on the third floor.
They wanted the building to be well-designed, be
competitive in the rental market and be a building with a
good residential environment.

위치 서울특별시 서초구 양재동 **용도** 제2종근린생활시설, 다가구주택 **대지면적** 253.60㎡
건축면적 151.10㎡ **연면적** 650.11㎡ **건폐율** 59.58% **용적률** 199.94% **규모** 지하 1층, 지상 5층
구조 철근콘크리트조 **시공** (주)예지인 종합건설

Location Yangjae-dong, Seocho-gu, Seoul **Use** Neighborhood facility, Multi-household house
Site area 253.60㎡ **Built area** 151.10㎡ **Total floor area** 650.11㎡ **Structure** Reinforced concrete

SITE PLAN

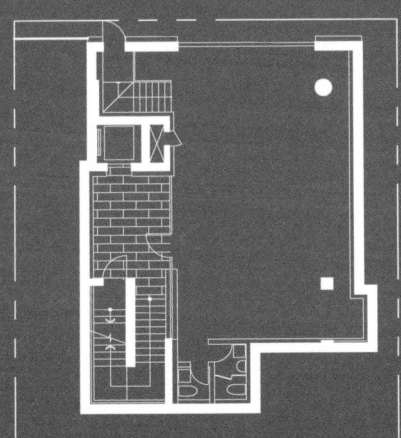
BASEMENT FLOOR PLAN

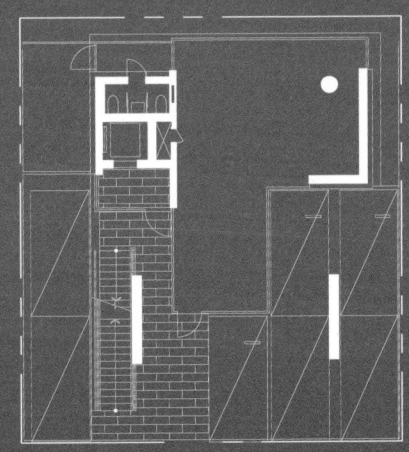
1st FLOOR PLAN

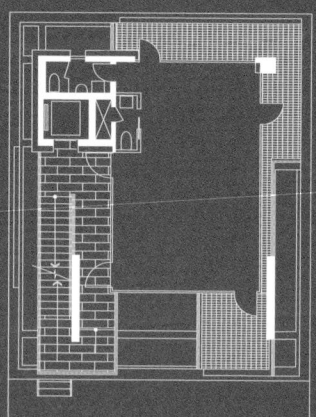
2nd FLOOR PLAN

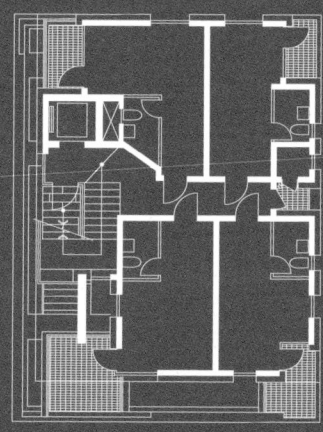
3rd FLOOR PLAN

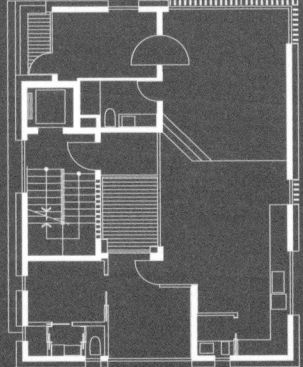
4th FLOOR PLAN

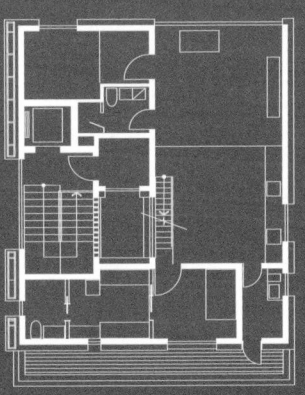
5th FLOOR PLAN

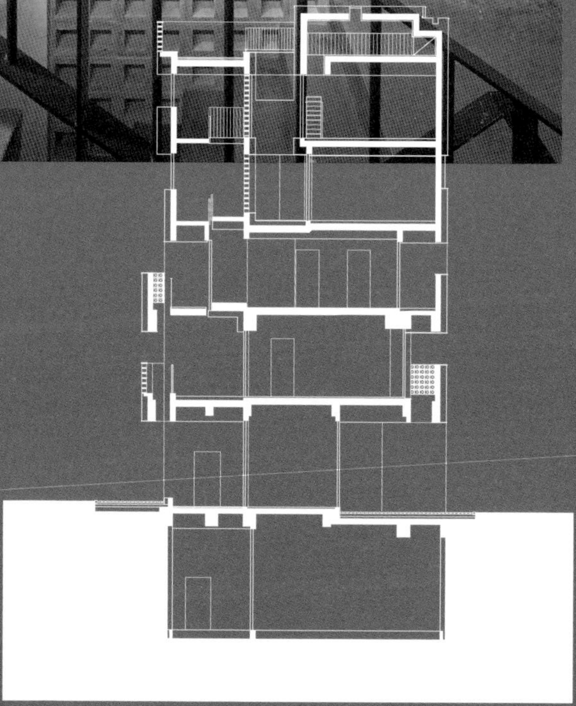

SECTION

강릉 Red Cube

강릉역 앞의 상업지역으로, 역에서 나왔을 때 바로 눈에 들어오는 필지이다. 역으로 필지에서는 강릉역과 그 주변을 넓게 조망할 수 있다. 20미터가 넘는 전면도로와 2미터가 채 안 되는 후면 골목길에 접해 있는 땅이었다. 땅은 55평의 크기로 주차 2대가 가능한 3개 층의 근린생활시설이 요구되었다. 1층은 주인이 직접 운영하는 식당으로, 2,3층은 매력적인 휴게 음식점이 임대하는 공간으로 구성하기를 원했다.

The site is seen as soon as one exits the station and enters the commercial district in front of Gangneung Station. From the site, one can widely see Gangneung Station and its surroundings. It was a land facing a front road that was more than 20 meters wide, as well as an alley that is less than 2 meters wide to the back. The land is 55 pyeongs large, and a three-story neighborhood living facility with two parking spaces was requested. They wanted to organize a restaurant run by the client on the first floor, and spaces to be rented by charming resting restaurants on the second to third floors.

위치 강원도 강릉시 교동 용도 제2종근린생활시설(일반음식점) 대지면적 183.00㎡ 건축면적 137.36㎡
연면적 293.23㎡ 건폐율 75.06% 용적률 160.24% 규모 지상 4층 구조 철근콘크리트조 시공 (주)동양산전

Location Gangneung-si, Gangwon-do **Use** Neighborhood facility **Site area** 183.00㎡
Built area 137.36㎡ **Total floor area** 293.23㎡ **Floor** 4F **Structure** Reinforced Concrete

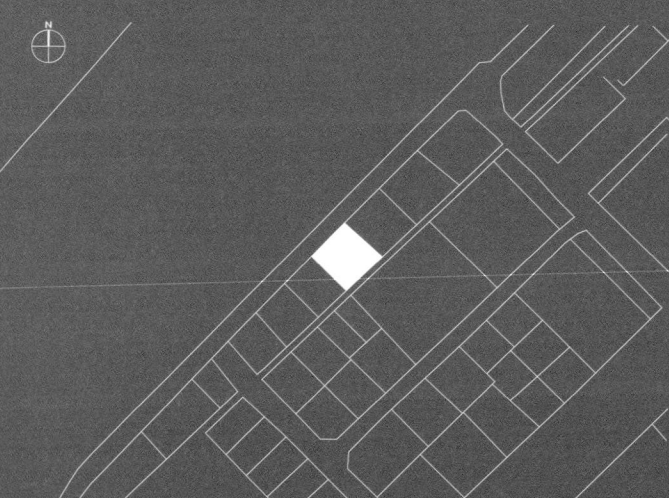

SITE PLAN

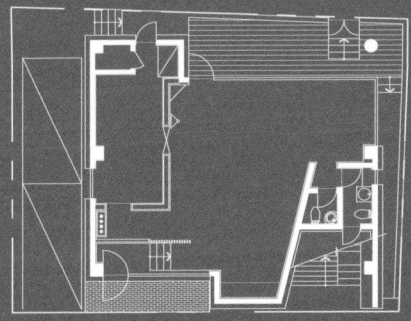

1st FLOOR PLAN

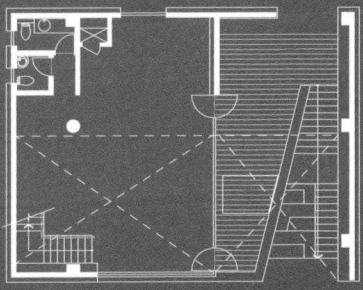

2nd FLOOR PLAN

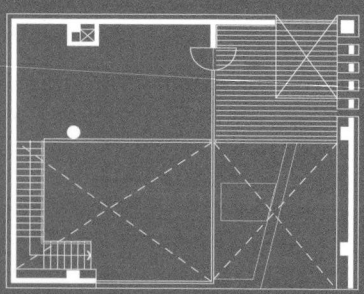

3rd FLOOR PLAN

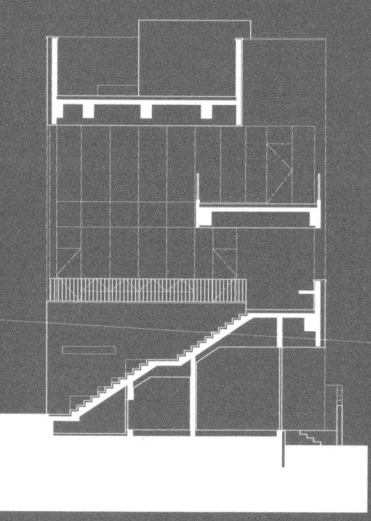

SECTION

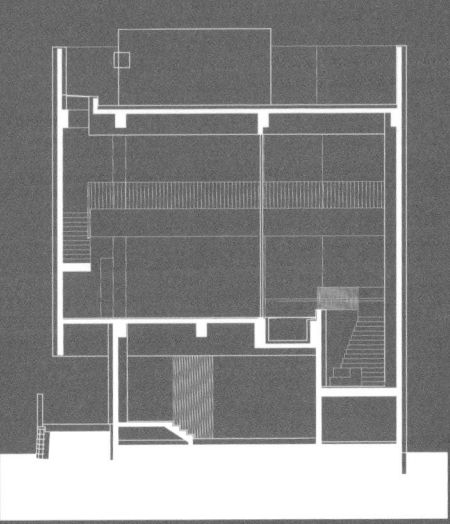

에세이

유형과 체계의 실험과 한계, 그리고 다른 가능성

현명석[*]

Essasy

Reconfiguring Typology and Systems of Practice: Richue Architecture's Experiments, Limits, and Other Possibilities

Myung Seok Hyun[*]

서울시립대 학부와 대학원에서 건축을 공부했다. 미국 조지아공대에서 20세기 중반 미국 건축사진을 이론화한 작업으로 박사학위를 받았다. 케네소주립대, 건국대, 경남대, 백석예술대, 서울시립대, 한양대에서 건축역사와 이론, 디자인 등을 가르쳤거나 가르친다. 『The Journal of Architecture』, 『The Journal of Space Syntax』, 『건축평단』, 『와이드AR』, 『Space』 등에 다수의 글과 논문을 실었다. 『건축사진의 비밀』(디북, 2019)의 공저자이며, 『건축표기체계: 상상, 도면, 건축이 서로를 지시하는 방식』(아키텍스트, 2020)을 엮었다. 서울에서 건축 매체와 재현, 시각성, 디지털 건축, 한국의 젊은 건축가들의 작업 등에 관한 연구와 저술에 몰두하고 있다.

Myung lives in Seoul, studying, teaching, and writing architecture. He received his PhD in architecture at the Georgia Institute of Technology, and his master's degree at the University of Seoul. Myung is an adjunct professor at Konkuk University and Hanyang University and has taught architectural history and theory and design studios both in the US and Korea. He is the editor and co-translator of Architectural Notation (2020) and the co-author of The Secret of Architectural Photography (2019), and has written for numerous other publications. His recent research interests include architectural medium and representation, the visualities of architecture, digital practice in architecture, and the works of young and upcoming Korean architects.

에세이   유형과 체계의 실험과 한계, 그리고 다른 가능성

1. 대안적 주거 건축의 조건

2010년대 초반 유행했던 이른바 '땅콩집'은 기존 단지형 아파트나 동네 '허가방'을 통해 복제, 양산되던 다세대/다가구주택의 진부한 유형을 대체하는 대안적 주거의 가능성을 보여줬다. 그러나 '땅콩집' 유행이 경제 모형의 가능성을 넘어 건축과 삶의 질을 높일 수 있는 더 큰 힘으로 이어지지 못한 점은 아쉽다. 여하튼 이후 떨어질 기미가 보이지 않는 높은 아파트값과 낮은 금리, 1인 가구 증가와 같은 가족 구조의 급진적 재편과 라이프스타일의 다양화로 요약할 수 있는 사회, 문화, 경제 조건의 결과 가운데 하나는 새로운 건축주 세대의 등장이다. 질 높은 주거 환경에 대한 욕망과 다양한 정보 채널을 통해 높아진 안목을 갖춘 이들 젊은 건축주 세대의 주택 수요는 최근 십여 년 동안 특히 한국의 젊은 건축가가 진입할 수 있는 시장을 형성했다. 예컨대 위층에 건축주의 집을 놓고 아래층에 작업실을 놓아 삶터와 일터를 결합하거나 임대형 원룸, 오피스, 상가 등을 놓아 안정적 임대 수입을 노리는 다세대/다가구주택, 상가주택, 소규모 전원주택 등이 이 시장을 이루는 주요 상품이다. 다채로운 작업 이력이 돋보이는 리슈건축

1. Recent Developments in the Market and the Clientele

The so-called "Peanut House," a budget-friendly duplex residential type that became quite popular in the early 2010s in Korea, demonstrated a viable way of financing and building a new form of urban and suburban living. To some, it was seen as a sound alternative to the colorless high-rise "apartments" developed by corporates and bureaucrats or the banal and outdated multi-unit/multi-family houses, replicated by the so-called "heo-ga-bang," the local pseudo-architects and the affiliated builders that literally copy/pasted existing plans and buildings for maximum FAR and hassle-free permission. However, it is a pity that the sensation, due to reasons, failed at reaching beyond a mere financing solution, and at evolving into a more critical architectural force that could mitigate the problems of urban housing in Korea at a more profound level.

In any case, high housing prices, low interest rates, and recent socio-cultural changes—lifestyles are increasingly diversified, while families are increasingly minimized—have led to the emergence of a younger generation of architectural clients. With a desire for unconventional and high-quality living environment and an informed and acquired taste for better design, the new clientele has formed a compact but lively market for young Korean architects to practice and realize their projects. Small-scale multi-unit/multi-

family housing or mixed-use commercial/residential buildings that guarantee steady income from lease or suburban units that offer enough recreational or workspace are those in demand here. Richue Architecture (led by Man-Sik Hong), with a list of quality projects in many sizes and programs under their belt, are one such firm.

Survival is never easy, especially for young and upcoming designers struggling in the market of Korean building industry. Lage-scale projects are reserved for corporates or renowned foreign/global firms, while ateliers led by the older generation of established star-architects are still finding ways to be competitive. All sorts of capitalist desires proliferate and swarm cunningly to maximize whatever they desire in whichever way they can. The new clientele, despite their tight budget, are getting increasingly smarter and knowledgeable. They are, on the other hand, generally reasonable and open-minded, with hopes to conduct better lives in better architecture. To survive, as well as to be respectful of the clients' good intentions, the architect simply cannot proclaim authority as an Author. Autonomy, at least in its critical sense, seemingly has no place here.

However, architects too have desires, which they cannot and should not simply abandoned. Formulated through training in prestigious school and professional practice, architects are indeed the agents in giving significance to the discipline.

(홍만식)은 특히 이런 시장에서 오랫동안 착실하게 경험을 쌓아온 건축 전문가다.

온갖 욕망이 정교하게 움직이고 교차하는 한국 건축 시장에서 살아남기는 절대 만만치 않다. 점점 똑똑해지는 건축주의 눈높이와 다양한 라이프스타일에 부응하는 서비스를 제공하지 못하면 도태된다. 빠듯한 예산이나마 좀 더 나은 주거 환경을 바라는 마음에 건축가를 찾은 건축주의 선한 의도 앞에서 어줍잖은 작가 행세나 건축 자율성autonomy의 논리는 별 소용이 없다. 그러나 건축가 역시 교육과 실무를 거치며 자연스럽게 형성된 자의식과 인식과 태도, 그리고 좋은 건축을 향한 욕망을 그냥 버릴 수는 없으며 버려서도 안 된다. 첨예한 자본의 논리와 건축가의 자의식, 생존의 필요성과 도시 리얼리티를 향한 비판의 당위성 사이 어느 곳에서 어떤 전략으로 움직일 것인가? 사실 따지고 보면 세부 조건이 조금 달라졌을 뿐, 이것은 긴 건축의 역사에서 그다지 새삼스러운 질문도 아니다. 질문은 도시와 그 도시 안에서 건축이 가져야 할 형태의 문제로 귀결되기도 한다. 이때 도시란 건축이 들어서는 바로 그 땅을 가리키는 것은 물론이요, 그곳에서 작동하는 수많은 힘의 복합체를 가리킨다. 물리 환경으로서 도시는 고정된 유한 자원이지만 그 속에서 온갖 욕망과 흐름과 규제는 끊임없이 재편된다. 재편의 양상은 불특정 다수의 변수에 따라 전개되는지라 논리적인 동시에 복잡하고 예측하기 어렵다. 건축가란, 어쩌면, 이런 복잡다단한 도시의 문제를 풀 수 있는 최선의 해답을 건축이라는 매체로 체화해 제안하는 전문가다.

2. 유형, 도시의 복잡성과 건축의 자율성 사이

도시는 끊임없이 변화하는 조건과 변수에 따라 재편된다. 예컨대 오늘날 여기저기에서 전가의 보도처럼 활용되는 이른바 '스마트도시'는 이런 재편의 주체를 사물 인터넷과 빅 데이터와 인공 지능으로 삼아 최적화의 속도를 높이고 자동화를 도모하는 기획이다. 반면 건축은 단단하고 고정된 형질에 가깝다. 상대적으로 변화에 둔하고 무겁다. 영속성이나 자율성을 중요한 가치로 삼는 건축 기율 안에서는 더 그럴 것이다. 그래서 도시와 건축, 둘

How should one navigate within this world of all-encompassing, accelerating, overwhelming, and the very real capitalist regime? How should one allocate oneself within the margins of reality yet somehow maneuver into the reality? In fact, although pronounced in different ways and dependent on slightly different conditions, such questions are hardly new in the history of architecture. They have manifested themselves as questions of autonomy, criticality, post-criticality, and such. Here, we will focus more on the question of typology, with which architects intervene within the city, a complex of diverse interests and forces. Capitalist desires, absurd regulations, and countless factors constantly re-assemble the city.

The patterns of such re-assemblage are unpredictable yet paradoxically logical.

2. Typology: Between Autonomy and Complexity

Cities constantly re-assemble according to the changing conditions and parameters. Some notable architectural / urban projects such as parametricism and the more recent "smart city" are attempts at optimizing and automating the processes of such re-assemblage. And because these projects rely on indefinite and generative relations of factors and performances, architecture is treated more as a trigger than a presence in itself. It is no secret that architecture is profoundly fixed, solid, and resistant to change. It endures, rather than submits. Thus architects, whose practice must

사이를 오가는 건축가에게는 반드시 딜레마가 주어진다. 끊임없이 재편되는 도시의 복잡성 안에서 건축은 어떻게 그 자율성을 유지할 것인가? 온갖 욕망이 충돌하는 시장 안에서 지켜내야 할 건축의 가치는 무엇이며 어떻게 그것을 지켜낼 것인가?

오랜 건축의 역사에서 이런 질문과 질문이 제기하는 도시와 건축의 실질적 문제에 대한 역사, 이론적 해답, 그리고 실천적 해결책이 됐던 것이 바로 '유형typology'이다. 유형은 곧 건축이 자율성을 가지며 도시의 복잡성에 유연한 대응하는 것을 가능케 하는 건축적 개념이자 실체다. 책에서 제시된 좌향, 여백, 표층과 같은 개념어도 실은 리슈건축이 오랫동안 천착했던 구체적 건축 유형, 예컨대 홍만식이 중요한 참조체로 삼는 전통건축, 예컨대 전남 구례 운조루나 도시형 한옥에 내재한 건축 체계를 현대 도시와 건축의 조건 안에서 다시 작동시키려는 기획의 산물이다. 예컨대 다양한 성격의 채가 각각 앉는 자리와 방향, 칸과 채의 분절을 통해 비워지는 마루와 마당의 다채로운 스케일과 위상, 또는 대문채나 행랑채의 두터운 켜가 형성하는 공간적 깊이와 입체적 체험 등은 전통건축의 속성이기도

always negotiate between diverse interests, are presented with a dilemma. How can architecture sustain its form, its autonomy, within the urban flux?

This question or its dilemma, persistent throughout the long history of architecture, is at the very core of the theory and the practice of typology—that is, the abstract and collective form of architecture induced from the empirical processes of making and observing cities, as well as that which prescribes and generates particular instances of the built. Therefore, typology is both the real and the virtual, which enables architecture to embody the complexities of its urbanity without losing its autonomous force. In fact, the notions of "adaptive reorientation," "the common space," and "thick surfaces" summarized in this book—some open-ended yet critical attempts at comprehending the design strategies of Richue Architecture—all constitute what may be a contemporary projection of a type. The type being projected here, if I were to specify, is that of traditional hanok (Korean house), notably its deep structure exemplified in Unjoru in Gurye, Jeollanam-do, or the urban hanok, which refer to the variants of traditional hanok that adapted to the new urban context during modernization in Korean cities. It should be noted that Man-Sik Hong, the founder of Richue Architecture, has studied in detail such types of traditional forms. Hong's agenda, if any, is to revitalize this deep form as a generative force that can still perform

its capacity through its creative use. For example, the traditional ways in which a chae, each with its own socio-cultural attribute, is placed and oriented in relation to the surroundings and the other chaes that constitute the complex of hanok, the various scales and shapes of void spaces resulting from flexible groupings of kans, or the visual depths created by the thick layers of daemunchae or haengrangchae are all typological properties that can still inspire viable design tactics in contemporary cities. Although the context has changed from the natural to the artificial, and thus the density and the regulating factors have also gone through a radical shift, these principles can and do apply in Richue's work.

3. Traditional Types Wrapped in Modernist Words

Apart from the occasional pixel-like openings, the Black Box in Anam-dong appears to be a monolithic cube of bricks coated with black paint. At first, despite its relatively small scale, the building looks solid, calm, and quite stubborn, rather resistant than engaging. However, a closer look immediately reveals that what appears to be a "brick monolith" is only a layer-thick skin that loosely wraps the inner guts of the building. The inside, maintaining a short gap from the brick skin, is further divided into two blocks. The blocks are raised (or lowered) by a half floor to one another in section—somewhat similar to the Corbusian split-level of Villa Carthage—,

하지만, 달라진 밀도와 규제의 복잡성 속에서 여전히 유효하며 얼마든지 보편타당한 현대 건축의 디자인 전략으로 작동할 수 있다. 리슈건축의 작업이 그 방증이다.

3. 전통건축 유형과 모더니즘 어휘

몇몇 개구부 말고는 흑색 도장 벽돌이 단단히 에워싼 채 좀처럼 내부를 드러내지 않는 안암동 「블랙 박스」의 첫인상은 차분하고 무겁고 불친절한, 단단한 덩어리다. 그러나 가까이에서 보면 검은 벽돌은 실은 덩어리가 아닌, 건물과 도시가 만나는 바로 그 면을 만드는 얇은 켜의 성분임이 곧 드러난다. 정작 내부는 이 켜에서 조금 물러나 좁은 사이-공간을 두고 층과 층이 교차하는 두 동의 건물로 다시 나뉜다. 44㎡의 좁은 건축면적 안에서도 얇거나 두꺼운 공간 켜를 겹겹이 에워싸 개방감과 깊이감을 구축했다. 다공 쌓기가 만드는 벽돌 외피의 입자화된 패턴은 변화하는 빛을 더 극적으로 예민하게 형상화해 내부 공간에 투사한다. 수직으로 열린 사이-공간으로 공간과 공간이 나뉘는 동시에 연결되는 모양새는 흡사 툇마루와 쪽마루를 거쳐 구들방과 마루를 오가는 그것을 닮았다. 한옥의 수평적

and between these blocks are the stairs and the lightwell that connect and mediate the carefully articulated spaces. The building thus succeeds in creating multiple layers of varying thicknesses and increased sense of depth, despite its compact building area of 44㎡. The particlized and porous construct of the brick exterior reacts to light quite sensitively, creating clear yet ephemeral patterns of cast shadow inside. Adolf Loos's Raumplan may come to mind, yet the Black Box resembles more the spatial relation of hanok, which cleverly connects the somewhat incompatible main spaces (such as maru and gudeulbang) through inconspicuously placed subspaces (toetmaru or jjokmaru): a more complex chain of space-to-subspace-to-space than the Loosian room-to-room relation. Here, the horizontal and loose complex of hanok reconfigures into a compact and articulate cube in a tight urban setting.

The Interacting Cube in Yonghyeon-dong, Incheon is relatively larger in scale than the Black Box. As the name suggests, the "interactions" and overlaps of solid / void, closed / open, interior / exterior are quite active, and offer an unusual look to the otherwise banal cityscape. The large chunks of space emptied and scooped out from the rectilinear mass remind us once more of the issue of urban density (more specifically FAR). Both the sights of the natural and the artificial enter through these empty spaces, and stay juxtaposed against the architecture. The building thus

에세이 유형과 체계의 실험과 한계, 그리고 다른 가능성

질서가 수직적 질서로 번안된 것이기도 하지만, 이는 동시에 '라움플랜Raumplan'을 통해 근대건축을 '공간들'의 양상으로 새롭게 정의하고자 했던 20세기 초 아돌프 로스Adolf Loos의 어휘이기도 하다.

「블랙 박스」에 비해 비교적 넉넉한 건축면적의 인천 용현동 「인터랙팅 큐브」에서 채워진 곳과 비워진 곳, 열림과 닫힘, 내부와 외부를 교차, 중첩해 개방감과 깊이감을 구축하는 전략은 더 자유로워 보인다. 직육면체 덩어리 이곳저곳을 명쾌하고 과감하게 비워낸 모습은 건축이 도시 속에서 가질 수 있는 밀도의 문제를 다시 생각하게 한다. 이들 공간을 통해 도시와 하늘의 풍경을 끌어들여 건축과 중첩하는, 특히 2층 카페의 시선에서 건물 내부의 채워진 곳과 비워진 곳, 더 나아가 가로와 도시를 함께 압착해 풍경으로 펼치는 솜씨가 인상적이다.

건축을 시각 기계로 보자면, 양재동 「보이드 라인」는 일련의 작업 가운데 가장 돋보인다. 「보이드 라인」의 특장점은 건물과 도시 가로 사이 경계를 다루는 방식에서 발현된다. 도시 가로, 특히 북쪽 전면 도로를 접하는 파사드의 경계면에서 층층이 분절돼 물러나거나 돌출되는 매스와 테라스의 '띠'의 깊이 변화, 그리고 이들 fully functions as an architectural frame, crafting a condensed and revitalized look of what lies beyond its boundaries.

If we were to consider architecture purely as a visual machine, the Void Line in Yangjae-dong is quite the surprise. The visual effectiveness is due to the way in which the architecture meets with the street — that is, how the boundaries between the interior and the exterior, the building and the street, or the building and the city are thickened, widened, and layered to embrace urban complexities. On the north façade facing the main street, the bold horizontal masses and strips that stack up to recede and protrude offer the viewer navigating on its edge a series of unexpected scenes. The masses, planes, and lines intersect each other aggressively. The building / machine is truly generative, in the sense that its systematic construct mounts to contingent events.

The Ground in Seongsu-dong displays its daunting presence in a banal yet rapidly changing urban context. The building combines diverse programs such as automobile exhibit, photography studio and workshop, and retail, and a multi-purpose venue for events, as well as living. The design incorporates glass boxes or empty spaces by lifting the mass above the ground or expanding gaps between masses. The result is an irregular and unconventional tripartite construct with three different voids, each with its own scale and properties. What triggers such a diversity

성수동「더 그라운드」

are the ways in which the boundaries are measured against the deformed site and the specific programmatic needs. The lower exhibition space (and parking), informed by the vector created by the acute angle of the plot, incorporates a conspicuous staircase (and seating) that dramatically connects the ground level to the third floor. The photographer's studio with its high ceiling is snugged inside the fourth and the fifth floors. Furthermore, the venue space located between the sixth and the seventh floors demonstrate, once again, the possibility of an alternative urban density that resists the status quo of its surroundings.

The bold shaping of voids, strong presence of stairs, and intersecting lines of sight across are also the significant themes appearing at the Red Cube in Gangneung. The Red Cube is still the more dynamic and playful, but its ground-level part, currently housing a small diner squatted under the proportionately large upper "head" of the building, acts as a medium that successfully negotiates a resolution between the polarizing aspects of the site. The low-profile ground level is itself a boundary, an open-ended one, which responds to both the front and the rear, the busy and commercial main street and the quiet rural residential area. The global nature of the transit station across and the locality of the specific context are ideally juxtaposed and represented here, a feature that speaks to the architect's talent and mastery of scale.

Essay　　　　　Reconfiguring Typology and Systems of Practice

띠 사이를 종횡으로 가로지르는 외부 계단은 경계 위, 그리고 경계 내부 곳곳에서 의외의 풍경을 선사한다. 도시와 건축, 가로와 건물, 밖과 안의 경계를 벌려 충분한 두께를 주고 수직과 수평과 대각의 덩어리와 면과 선이 시각적 장치로서 어지럽게 교차하는 가운데 다양한 시점의 가능성을 여는 동시에 시차를 발생시킨다. 최근 완공된 성수동 「더 그라운드」는 덩어리를 수직으로 높이 띄우고 유리 상자를 삽입하거나 덩어리가 나눠진 틈을 벌려 과감히 빈 곳을 더한 모양새다. 결과적으로 저층부, 중층부, 상층부에 각각 다른 보이드가 형성되는데, 특히 이들 사이에는 필지가 가로와 비스듬히 만나며 형성되는 둔각과 예각의 세 면을 처리하는 방식과 연계된 프로그램에 따라 차이가 촉발된다. 예각의 방향성을 그대로 받아 지상층 가로에서 3층까지 연결하는 계단을 입체화하고 이것을 따라 흐르는 동선과 시선을 주차장과 영리하게 연계한 저층부 전시 공간, 4층과 5층 내부에 사진작가인 건축주의 촬영을 위해 마련한 중층의 작업 공간, 외부에서 도시와 시선을 교환하며 다양한 이벤트를 수용할 수 있는 6층과 7층의 다목적 공간은 도시의 밀도에 적응하며 얻을 수 있는 보이드의 가능성을 보여준다.

「더 그라운드」에서 선보인 과감한 보이드 만들기와 진입 계단의 입체화, 도시와 건축의 시선 교환과 같은 주제는 강릉 「레드 큐브」에서도 돋보인다. 반면 건물 전면과 후면의 달라지는 높낮이와 맥락을 담담하게 받아 디자인 변수로 활용한 지상층의 식당 공간에서는 세심하게 스케일과 공간을 다루는 건축가의 솜씨를 엿볼 수 있다. 강릉 역세권의 광역global 스케일과 전/후면 가로의 단차 덕분에 더 낮아진 고만고만한 높이의 지역local 스케일, 그리고 그에 따라 사뭇 달라지는 위요와 재현의 감각들이 무리 없이 조합된 수작이다.

4. 일상의 익숙함을 넘어서는 생경함

도시는 높은 복잡도의 '환경environment'에 가깝다. 반면 그 끝 모르는 복잡도를 역사적이고 경험적인 실체로 한정시켜 안정적 강도intensity로 잡아두는 건축 유형은 하나의 '체계system'라 부를 수 있다. 따라서 체계, 곧

4. The Unfamiliar Beyond the Banal

The city is an environment, a highly complex one. The type, on the other hand, is a system that confines and stabilizes such complexities into a historical and empirical reality. Therefore, the complexities of a type / system can never surpass those of the city / environment, especially when the former is secured by the conventional. In this sense, the diagram, a conceptual and pragmatic design instrument that has gained architectural significance since the Deleuzian invasion, may also be understood as a kind of system. The diagram, however, may do without the historical or conventional implications, and only tend to the realities of things and surpass them precisely by registering their absurdities. The diagram, as such, becomes an alternative to the seemingly outdated and unproductive historical typology, amidst globalization and neo-liberalism that leave little room for adherence to the past.

I mention these notions of typology and diagram as to refer to their stabilizing—and projective, when successful—capabilities. These are architectural instruments, and Richue's work does show the symptoms of being both typological and diagrammatic. Although typology-based, Richue's process is curiously pragmatic, in the sense that it carefully registers the site-specific and programmatic data and regulations as to reconfigure them into a well-crafted modernist form. If I were to pick from

강릉 「레미큐브」

유형의 복잡도는 언제나 환경, 곧 도시의 복잡도를 넘어설 수 없으며, 그것이 건축 선례와 관습에 묶여 있다면 더 그럴 것이다. 질 들뢰즈Gilles Deleuze의 생각이 건축 담론 안에서 변용된 후 중요한 개념적, 실천적 기제로 등장한 다이어그램은, 어찌 보면, 이와 같은 건축 유형을 대체했던 또 다른 체계다. 유럽 역사 도시, 그리고 그 속에서 안정적으로 자율성을 보장받던 건축 유형이 가질 수밖에 없는 한계를 넘어 다이어그램은 1990년대 이후 세계화와 신-자유주의 이후 급증하는 고밀도 거대도시 환경의 높아진 복잡도를 수용할 수 있는 대안으로 작동해왔다. 앞으로 빅 데이터를 기반으로 하는 인공 지능이 또 다른 층위의 복잡도를 수용할 수 있는 대안적 디자인 체계로 작동할 수 있을지는 두고 볼 일이다.

굳이 이 이야기를 꺼내는 까닭은 그것이 도시의 복잡성을 건축 매체를 통해 인식, 수용하고 다시 건축화하는 방식에 대한 이야기라는 데 있다. 리슈건축의 작업에서 그 방식은 주로 건축 유형의 참조와 변용으로 나타난다. 물론 다이어그램의 흔적도 엿보인다. 그리고 실제로 리슈건축의 건축적 해결책, 곧 건물로서의 결과물 또한 그리 나쁘지 않다. 여기 소개되는 다섯 개의 근작 가운데 굳이 꼽자면, 이런 측면에서 가장 돋보이는 작업은 「보이드 라인」이다. 두툼한 경계면 위에서 정통의orthodox 건축 어휘를 통해 자신 있게 펼쳐내는 시각적 경험은 실로 우발적이고 예측하기 어렵다. 그 복잡도가 안정적 유형의 차원과 한계를 넘어섰다는 말이다.

그러나 아쉬움은 여전히 남는다. 유형이든 다이어그램이든, 또는 수화digitization나 이산성 discreteness을 소극적으로 형상화한 것 이상으로 보기 어려운 몇몇 디자인 제스처이든, 냉정하게 말해 리슈건축의 작업에서는 어떤 급진적 양상을 찾기 어렵다. 대체로 절충적이란 말이다. 글 첫머리에서 언급한 건축주와 건축가 모두의 생존이 첨예할 수밖에 없는 리얼리티의 한계 탓일 수도 있고, 고형의 실체로서 전통 건축의 유형이 갖는 태생적 한계 탓일 수도 있다(굳이

the works introduced here, the Void Line is perhaps the most notable case, as its systematic inclusion of conventions and data surely evolve into somewhat explosive yet manageable indeterminism. Richue's orthodox modernist language, boldly spoken through masses and spaces, creates thick boundaries and surfaces that offer unpredictable visual experience. The inherent complexity of the Void Line surely surpasses that of its typological origin.

Although with such admirable qualities, Richue's architecture generally avoids being radical. Some clever design gestures are found here and there, but they usually fall short of inciting truly avant-gardist moments. Their work rarely pushes the envelope to its maximum or to the point of strangeness and absurdity. This is perhaps because of the economic condition the firm must face (as I have briefly outlined above) or the irresistible nature of regression, embedded within the very forms of traditional architecture that Richue's architecture often refers to. In short, Richue's practice remains to be a compromise, safely managing relations with all encounters yet never really exploiting any. I do hope, however, to see more absurdity, the uncanny or weirdness in Richue's future work—typologies and diagrams exploited and expressed in maximum volume, as to present the architecture audience with something truly novel. The hope is not baseless, as symptoms are plentiful. Richue's practice

따지자면, 아마도 후자 탓이 더 클 것이다). 그러나 유형이라면 유형으로서, 다이어그램이라면 다이어그램으로서, 아니면 또 다른 무엇으로서, 그 체계가 도시의 복잡도를 최대치로 드러내는 생경한 풍경을 리슈건축의 작업에서 볼 수 있으면 좋겠다. 이런 기대는 말한 대로 아쉬움을 전제로 하지만, 동시에 리슈건축의 작업 곳곳에서 언뜻언뜻 목격할 수 있는 과감하고도 우직한 형태와 공간 짓기 구사력에서 비롯하는 것이기도 하다. 급진성과 생경함의 '징후'가 충분하다는 말이다. 실제로 비슷한 조건에 처한, 그러나 조금 더 젊은 건축가 세대의 작업에서 이만큼 형태나 공간을 건축의 중요한 주제로 삼는 건축가는 그리 많지 않다. 좋은 건축을 넘어 도시의 힘들을 날것 그대로의 모습으로 드러내는 급진적 건축을 리슈건축의 형태와 공간에서 볼 수 있을까. 성공 여부는 건축을 결정짓는 도시의 변수들에 내재한 다양한 가능성을 어떤 변곡inflection의 지점까지 가져가 실험할 수 있는지에 달려있다.

is smart enough and skillful enough, as well as collaborative and knowledge-based in nature, even when compared to much younger upcoming architects in the scene. Moreover, the orthodox language of form, space, and the tangible are still meaningful and present in their work and approaches, at least in form, the early modernist projects of the last century. Can Richue go beyond their limits and propose visceral entities that embody the raw forces of the city—so raw that it hurts to know? What must occur is an inflection that reconfigures the system, the typology or the diagram. Richue have devoted themselves in learning and adopting such systems to a respectable degree. Let us see if they can exceed them.

부록

리슈건축 프로젝트 2012-2020

Appendix

Richue Architecture's Works 2012-2020

1　　　　　　　2　　　　　　　3　　　　　　　4

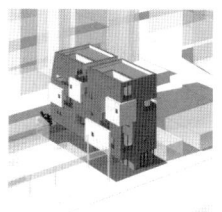   

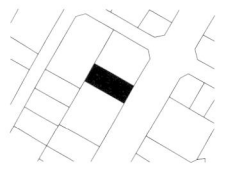 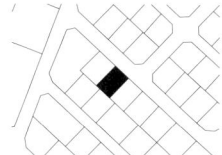 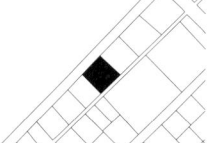 

구분	프로젝트	위치
1	Terrace Scape	울산광역시 중구 성안동
2	가락동 소소헌	서울특별시 송파구 가락동
3	강릉시 교동 Red Cube	강원도 강릉시 교동
4	거여동 Street Garden House	서울특별시 송파구 거여동
5	고양 삼송동 삼각창집	경기도 고양시 삼송동
6	고양 화정동 삼각집	경기도 고양시 화정동
7	동교동 UFO	서울특별시 마포구 동교동
8	망원동 에코하우스	서울특별시 마포구 망원동

부록　　　　　리슈건축 프로젝트 2012–2020

5	6	7	8

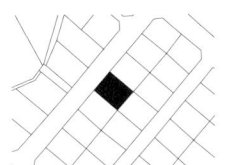	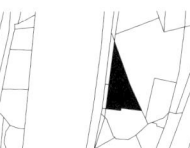		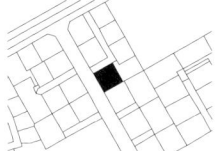

* 지적도의 하단이 남쪽을 향하도록 배치

대지면적	건축면적	연면적
316.60㎡	185.95㎡	594.22㎡
170.50㎡	101.40㎡	340.74㎡
183.00㎡	137.51㎡	316.16㎡
247.50㎡	159.87㎡	547.04㎡
262.20㎡	156.78㎡	469.38㎡
380.00㎡	222.56㎡	416.84㎡
235.10㎡	128.70㎡	689.10㎡
167.60㎡	95.23㎡	334.87㎡

9	10	11	12

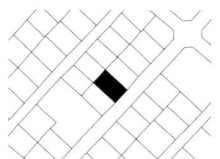	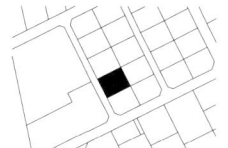	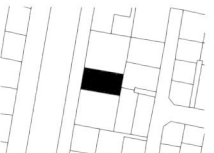	

구분	프로젝트	위치
9	문정동 상가주택	서울특별시 송파구 문정동
10	방배동 Yellow House	서울특별시 서초구 방배동
11	부천 상동 Screen Building	경기도 부천시 원미구 상동
12	북가좌동 다세대주택	서울특별시 서대문구 북가좌동
13	상도동 삼각집	서울특별시 동작구 상도동
14	상도동 세자매집	서울특별시 동작구 상도동
15	상도동 협소주택	서울특별시 동작구 상도동
16	서교동 고깔지붕집	서울특별시 마포구 서교동

Appendix　　Richue Architecture's Works 2012 – 2020

| 13 | 14 | 15 | 16 |

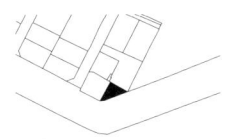 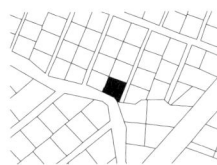  

* 지적도의 하단이 남쪽을 향하도록 배치

대지면적	건축면적	연면적
185.50 m²	111.23 m²	367.56 m²
195.00 m²	116.25 m²	485.44 m²
302.00 m²	165.80 m²	827.78 m²
336.00 m²	189.39 m²	660.00 m²
81.40 m²	48.83 m²	160.30 m²
119.00 m²	70.98 m²	195.87 m²
60.60 m²	36.13 m²	142.84 m²
154.00 m²	91.14 m²	304.85 m²

17　　　　　　　18　　　　　　　19　　　　　　　20

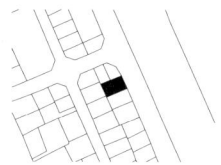 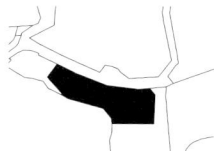  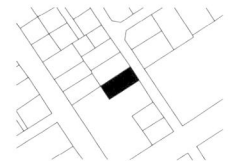

구분	프로젝트	위치
17	성북천변 계단집	서울특별시 성북구 보문동
18	신봉동 베네우스	경기도 용인시 신봉동
19	안암동 Black Box	서울특별시 성북구 안암동
20	안양시 붉은 벽돌집	경기도 안양시 만안구 안양동
21	양재동 Void Line	서울특별시 강남구 양재동
22	양주 봉양동 상가주택	경기도 양주시 봉양동
23	용현동 Interacting Cube	인천광역시 미추홀구 용현동
24	청라 커넬큐브	인천광역시 서구 연희동

| 21 | 22 | 23 | 24 |

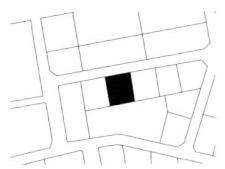 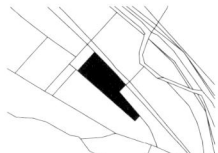 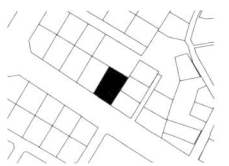 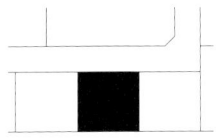

* 지적도의 하단이 남쪽을 향하도록 배치

대지면적	건축면적	연면적
210.90㎡	126.04㎡	545.81㎡
533.00㎡	210.98㎡	532.72㎡
77.40㎡	44.41㎡	127.14㎡
197.00㎡	118.14㎡	352.48㎡
253.60㎡	151.12㎡	650.71㎡
464.00㎡	91.80㎡	249.34㎡
240.40㎡	143.21㎡	567.35㎡
1260.00㎡	755.58㎡	8132.48㎡

25 26 27 28

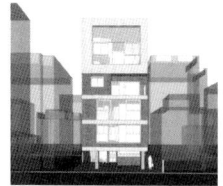

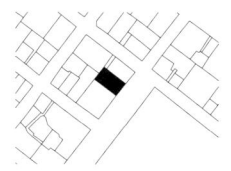
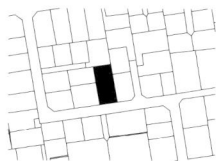
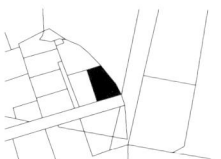
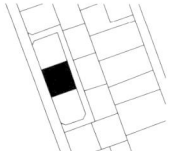

구분	프로젝트	위치
25	한강로2가 Y빌딩	서울특별시 용산구 한강로2가
26	화곡동 다세대주택	서울특별시 강서구 화곡동
27	통영 도마집	경상남도 통영시 광도면 죽림리
28	Scape 2712	서울특별시 강남구 삼성동
29	당진 층층마당집	충청남도 당진시 채운동
30	동패동 삼 지붕집	경기도 파주시 동패동
31	망원동 모퉁이집	서울특별시 마포구 망원동
32	목동 파노라마 하우스	서울특별시 양천구 목동

Appendix Richue Architecture's Works 2012−2020

29	30	31	32

	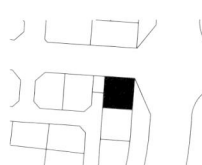		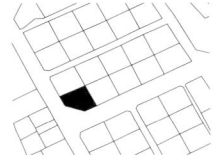

* 지적도의 하단이 남쪽을 향하도록 배치

대지면적	건축면적	연면적
146.80㎡	80.86㎡	332.20㎡
237.40㎡	137.74㎡	474.54㎡
340.00㎡	169.17㎡	413.07㎡
268.51㎡	161.11㎡	869.95㎡
262.00㎡	155.00㎡	565.25㎡
276.20㎡	165.62㎡	347.58㎡
192.50㎡	110.14㎡	482.39㎡
188.00㎡	111.62㎡	316.43㎡

33　　　　　　　34　　　　　　　35　　　　　　　36

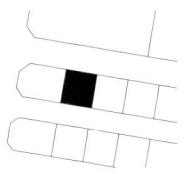
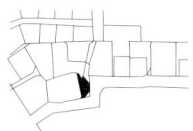
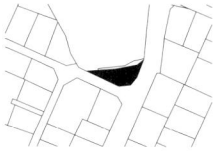
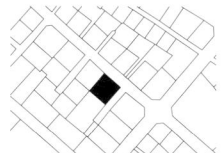

구분	프로젝트	위치
33	부천 중동 Void Box	경기도 부천시 중동
34	상도동 모서리집	서울특별시 동작구 상도동
35	상도동 반달집	서울특별시 동작구 상도동
36	수유동 모퉁이 집	서울특별시 강북구 수유동
37	신당동 모퉁이 빌딩	서울특별시 중구 신당동
38	신사동 가로수길 Void Line	서울특별시 강남구 신사동
39	안동 깊이 파인 집	경상북도 안동시 갈전리
40	양주 옥정신도시 상가주택	경기도 양주시 옥정동

부록　　　　　리슈건축 프로젝트 2012 – 2020

37 38 39 40

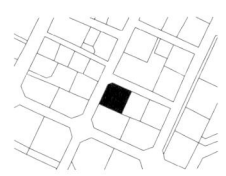  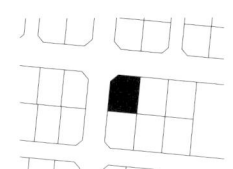 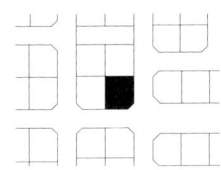

* 지적도의 하단이 남쪽을 향하도록 배치

대지면적	건축면적	연면적
351.20㎡	307.98㎡	704.73㎡
65.08㎡	34.44㎡	117.28㎡
175.00㎡	97.97㎡	287.18㎡
171.60㎡	96.08㎡	340.52㎡
212.00㎡	120.32㎡	524.26㎡
187.90㎡	111.12㎡	374.72㎡
366.20㎡	198.25㎡	493.91㎡
325.00㎡	188.72㎡	537.27

41　　　　　　42　　　　　　43　　　　　　44

구분	프로젝트	위치
41	용산 리슈빌딩	서울특별시 용산구 한강로2가
42	진주시 초전동 상가주택	경상남도 진주시 초진동
43	하남 위례 상가주택	경기도 하남시 학암동
44	한강로2가 Black Box	서울특별시 용산구 한강로2가
45	헬로우닥터 사옥	경기도 성남시 중원구 하대원동
46	부천 오정구 근린생활시설	부천시 오정구 오정동
47	상암 DMC H사 사옥	서울특별시 마포구 상암동
48	성산동 Urban Membrane	서울특별시 마포구 성산동

Appendix　　Richue Architecture's Works 2012 – 2020

| 45 | 46 | 47 | 48 |

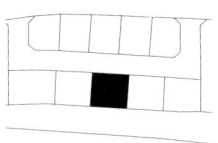 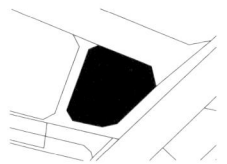  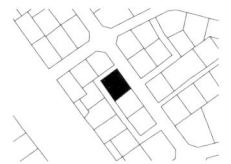

* 지적도의 하단이 남쪽을 향하도록 배치

대지면적	건축면적	연면적
298.50㎡	178.00㎡	712.20㎡
251.00㎡	149.91㎡	433.89㎡
283.00㎡	168.50㎡	499.97㎡
487.90㎡	288.44㎡	996.26㎡
429.30㎡	254.06㎡	900.68㎡
1,639.30㎡	1,115.00㎡	8,731.00㎡
3,172.00㎡	1,898.06㎡	38,683.40㎡
165.67㎡	81.33㎡	413.31㎡

49　　　　　　　50　　　　　　　51　　　　　　　52

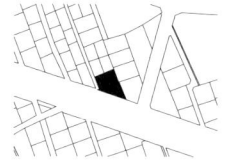
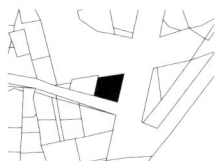
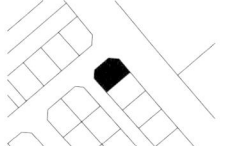

구분	프로젝트	위치
49	성수동 The Ground	서울특별시 성동구 성수동2가
50	울진 젠가 하우스	경상북도 울진군 울진읍 읍내리
51	하남미사 망월동 상가주택	경기도 하남시 망월동
52	냉천동 다세대 주택	서울특별시 서대문구 냉천동
53	성북동 다가구주택	서울특별시 성북구 성북동
54	천안 불당동 상가주택	충청남도 천안시 불당동

53 54

 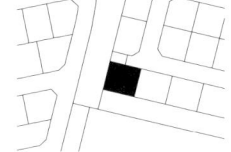

* 지적도의 하단이 남쪽을 향하도록 배치

대지면적	건축면적	연면적
190.18㎡	113.17㎡	823.39㎡
205.00㎡	117.05㎡	279.13㎡
250.20㎡	149.90㎡	493.99㎡
285.30㎡	169.54㎡	566.19㎡
175.28㎡	93.88㎡	209.55㎡
317.50㎡	190.16㎡	532.23㎡

좌향, 여백, 표층

이 책에 실린 내용은 월간 『건축문화』의 2020년 12월호의 특집 지면을 바탕으로 일부 수정된 것임을 밝힙니다.

초판 1쇄 인쇄. 2021년 5월 13일
초판 1쇄 발행. 2021년 5월 14일

기획. (주)리슈건축
편집. 정평진
디자인. 김범준
사진. 김재윤, 이한울
인쇄. 충주문화사

발행처. 도서출판 우리북 (대표 김영덕)
출판등록. 2010년 8월 28일 (제 321-2010-000175호)

06778
서울시 서초구 양재동 247
전화. 02-3463-2130
팩스. 02-2360-2150

ooribook.com

ISBN 979-11-85164-37-3
값 15,000원

어커먼즈 프레스는 도서출판 우리북의 임프린트 입니다. 이 책의 판권은 도서출판 우리북에 있습니다.

이 책의 내용은 저작권법에 의하여 보호받는 저작물이므로 본사의 허락 없이는 어떠한 형태로도 이용하실 수 없습니다.

이 책의 제목과 본문은 '을유1945' 서체를 사용했습니다.